BMW
5 Series

Other Titles in the Crowood AutoClassics Series

AC Cobra	Brian Laban
Alfa Romeo Spider	John Tipler
Aston Martin DB4, DB5 and DB6	Jonathan Wood
Aston Martin and Lagonda V-Engined Cars	David G Styles
BMW M Series	Alan Henry
Carbodies	Bill Munro
Citroën DS	Jon Pressnell
Datsun Z Series	David G Styles
Ferrari Dino	Anthony Curtis
Jaguar E-Type	Jonathan Wood
Jaguar Mk1 and 2	James Taylor
Jaguar S-Type and 420	James Taylor
Jaguar XJ-Series	Graham Robson
Jaguar XJ-S	Graham Robson
Jaguar XK Series	Jeremy Boyce
Lamborghini Countach	Peter Dron
Land Rover	John Tipler
Lotus and Caterham Seven	John Tipler
Lotus Elan	Mike Taylor
Lotus Esprit	Jeremy Walton
Mercedes SL Series	Brian Laban
MGA	David G Styles
MGB	Brian Laban
MG T-Series	Graham Robson
Mini	James Rupert
Morris Minor	Ray Newell
Porsche 356	David G Styles
Porsche 911	David Vivian
Porsche 924/928/944/968	David Vivian
Range Rover	James Taylor and Nick Dimbleby
Rover P4	James Taylor
Rover P5 & P5B	James Taylor
Rover SD1	Karen Pender
Sprites and Midgets	Anders Ditlev Clausager
Sunbeam Alpine and Tiger	Graham Robson
Triumph 2000 and 2.5PI	Graham Robson
Triumph Herald and Vitesse	Graham Robson
Triumph TRs	Graham Robson
TVR	John Tipler
VW Beetle	Robert Davies
VW Golf	James Ruppert
VW Transporter	Laurence Meredith

BMW 5 Series

The Complete Story

James Taylor

First published in 1999 by
The Crowood Press Ltd
Ramsbury, Marlborough
Wiltshire SN8 2HR

© James Taylor 1999

All rights reserved. No part of this publication may be reproduced or transmitted in any form or by any means, electronic or mechanical, including photocopy, recording, or any information storage and retrieval system, without permission in writing from the publishers.

British Library Cataloguing-in-Publication Data
A catalogue record for this book is available from the British Library.

ISBN 1 86126 178 0

Set in New Century Schoolbook

Produced by Focus Publishing, Sevenoaks, Kent

Printed and bound in Great Britain by The Bath Press

Contents

	Acknowledgements	6
	Introduction	7
	Evolution	8
1	Pedigree: BMW Before the 5 Series	9
2	First Generation: The E12 Models, 1972–81	22
3	Second Generation: The E28 Models, 1981–87	53
4	The E28 Super-Saloons	84
5	Third Generation: The E34 Models, 1988–96	97
6	The E34 Super-Saloons and Tourings	128
7	Fourth Generation: The E39 Models From 1995	142
8	The E39 Super-Saloons and Tourings	165
9	Buying a 5 Series BMW	176
	Index	190

Acknowledgements

While putting this book together, I called on the knowledge and expertise of many people, and probably tried the patience of a good number of them. They can't be blamed for any errors which have crept into my interpretation of what they told me about the 5 series BMWs.

Special thanks for help go to Alan Arguile at BMW South Africa, Rob Mitchell at BMW of North America, Alun Parry at BMW (GB) and to Herr Klebl and his staff at BMW Mobile Tradition in Munich. I am also grateful to Birds (UK Hartge representatives) and to Sytner of Nottingham (UK Alpina representatives).

Extra help with pictures came from Chris Horton (1986 E28 M535i), from Ian Kuah (E39 Hamann H5/450 V12), from Andy Tipping (E39 B10 V8), and from Dave Shepherd (E34 B10 Bi-Turbo). And last, but by no means least, I am grateful to Charles Armstrong-Wilson, editor of *BMW Car* magazine. His help ranged from the opportunity to make articles out of parts of this book, to help in sourcing photographs.

James Taylor

Introduction

The 5 series BMWs are such a common sight on our roads today that it's easy to forget their influence on saloon car development in the last quarter of the 20th century.

They have become the standard by which other family saloons are judged, and from the humblest 518i all the way up to the supercar M5, they offer an unrivalled motoring experience.

I hope this book will go some way towards explaining what lies behind the 5 series of today, and that it will make clear why all these cars are very much more than the family saloons they sometimes appear to be. Without them, the BMW company would be very different today – and so would the products of rival manufacturers who strive to emulate their achievements. I have deliberately gone into a lot of detail on every model, but to list every single option and variant in every single territory would take a book much larger than this one. I hope this will satisfy readers ... at least, for now.

Evolution – BMW'S E Series Codes

Modern BMWs – and that includes all of the 5 Series ranges – have a factory type code beginning with E. This E probably stands for Entwurf (design) and is allocated at the design stage; the numerical order therefore relates to the date when design commenced, and not to the date when a model entered production. Not every project has made it past the design stage, and there are therefore many gaps in the sequence of numbers associated with the production BMWs.

The first E series codes were allocated in the mid-1960s. What follows is a list of the main production models since then, with the 5 Series cars highlighted in bold type.

E3	2500, 2800 and subsequent big saloons of this range, 1968–77.
E6	Facelifted 02 models, 1973–75. (The original 02s did not have an E code).
E9	Six-cylinder coupés from 2500CS to 3.0CSL, 1968–71.
E10	2002 turbo, variant of the E6 02 models, 1973–75.
E12	**First-generation 5 Series saloons, 1972–81.**
E21	First-generation 3 Series, 1975–82.
E23	First-generation 7 Series saloons, 1977–86.
E24	6 series coupés, 1976–89.
E26	M1 mid-engined supercar, 1978–80.
E28	**Second-generation 5 Series, 1981–88.**
E30	Second-generation 3 Series, 1983–91.
E31	8 series coupés, 1989 on.
E32	Second-generation 7 Series, 1986–94.
E34	**Third-generation 5 Series saloons, 1988–96.**
E36	Third-generation 3 Series, 1991 on.
E38	Third-generation 7 Series, 1994 on.
E39	**Fourth-generation 5 Series saloons, 1995 on.**
E46	Fourth-generation 3 Series, 1997 on.

1 Pedigree: BMW Before the 5 Series

In every country where it is sold, the BMW 5 Series is recognized as the standard-setting medium-sized saloon, as the best car in its class. It offers ride comfort for five people and excellent reliability and durability, and can provide a wonderfully involving driving experience if its driver so wishes. Its variants spread right across the medium-sized range, from an entry-level 1.8-litre four-cylinder model to a luxuriously-equipped 4.4-litre V8 variant and a stunningly powerful 4.9-litre sporting model, taking in turbodiesels and estate-bodied types on the way. In short, the BMW 5 Series has all the bases covered.

There have been 5 Series BMWs for more than a quarter of a century now, and for most of that period the cars have occupied the same enviable position in the medium-sized saloon market. Yet although the cars have always had about them a certain conservatism so necessary for success in their chosen market sector, they have also embodied technical advances which have kept them ahead of their competitors. Evolution rather than revolution has always been the keynote of

The mainstream BMWs of the 1950s were the big Baroque Angel saloons, of which a 502 V8 is seen here. They were well-respected, but did not sell well...

Pedigree: BMW Before the 5 Series

... partly because their styling harked back to BMW's golden age of the 1930s, of which this 326 cabriolet is typical.

their development, and while the fourth-generation cars of the later 1990s are very different indeed from the first-generation models of the early 1970s, it is possible to see in them an evolution of the same philosophy.

BMW got it nearly right first time and has been perfecting the concept ever since. And – even more remarkably – the company has managed to broaden its customer base quite dramatically while remaining true to the original concept of the 5 Series. The first cars were unashamedly designed for enthusiast drivers, and in this way BMW managed to cream off a worthwhile percentage of the medium-saloon market which was denied to Mercedes-Benz. It was these enthusiast drivers who continued to form the core of the customer base for the next quarter of a century, while the 5 Series was carefully developed into a car which appealed right across the spectrum of buyers in the medium-sized saloon class. The car's development from a respected niche model into the standard-setter for the whole class was a masterpiece of engineering development and marketing, and demonstrates vividly both the depth of talent and the management vision that have made BMW into one of the world's leading car manufacturers.

Pedigree: BMW Before the 5 Series

The Isetta

BMW built the Isetta bubble-car under licence from Iso S.p.a. of Italy, who were primarily makers of motor scooters and three-wheeler utilities. Iso introduced the Isetta in 1953, and also sold a licence for its production to Velam in France. Iso's owner, Count Renzo Rivolta, eventually spent the profits from these agreements on making the Euro-American Iso-Rivolta and Iso Grifo supercars.

The BMW Isetta 250 dispensed with Iso's two-stroke engine, using instead the four-stroke 247cc single-cylinder from the R25 motorcycle. From February 1956, there was a companion-model Isetta 300, with the more powerful 297cc engine from the R27 motorcycle. Other changes included smaller headlamp cowls after 1955 and a completely revised glass area with larger side windows from October 1956.

The original Iso car, the BMW version, and the Velam all had twin rear wheels, but a version of the Isetta 300 built under licence from BMW in Britain from 1958 actually had a single rear wheel, because three-wheelers attracted less Purchase Tax and their road fund licence was cheaper! Just 1,750 three-wheelers were built.

In the mid-1950s, the Isetta cost just 20 per cent as much as the cheapest of the Baroque Angel 501 saloons.

The bulky styling was still present in the 505 limousine, drawn up in 1955 but never put into production. One of the only two built is seen on a rare outing in West German Government service.

Pedigree: BMW Before the 5 Series

THE POST-WAR PERIOD

The company's expansion during the period of the 5 Series' existence has been the envy of its rivals, but things were not always so rosy for BMW. Only a dozen years before the first 5 Series was introduced, BMW had been in very bad shape indeed.

The problem was that the company had never really found its niche in the car market after the end of the Second World War in 1945. Initially, it struggled to survive in the harsh post-war economic climate, having lost its main factories which were situated in the Soviet-occupied sector of Germany. Motorcycle production started hesitantly in 1948, and several prototype cars were built in the late 1940s, but it was 1951 before the company was prepared to show a new car in public, and 1952 before that car entered volume production.

Unfortunately, the new 501 harked back too much to BMW's successful designs of the 1930s, with the result that its styling dated very quickly and the cars earned the appropriate but rather disdainful nickname of 'Baroque Angels'. The 501 started off with a revival of the much-copied pre-war 2-litre six-cylinder engine, but it was a heavy and cumbersome machine and even a succession of 2.6-litre and 3.2-litre V8s could not turn it into a really credible contender in the marketplace. Attempts to find export sales in the USA with sports and grand touring models based on the Baroque Angels' running-gear were none too successful, either.

The running-gear of the Baroque Angels was also used for more glamorous machines, like this 503 coupé. Sales were too small to make the profits BMW needed.

Pedigree: BMW Before the 5 Series

The big V8 engine was also used to power the gorgeous 507, BMW's answer to the Mercedes-Benz 300SL. Again, sales proved disappointing.

The Baroque Angels

The 500-series cars may not have been BMW's most glamorous products, but these big, sturdy middle-class machines were the mainstay of the company's car division from 1951 to 1964. They were nicknamed 'Barockengel' – Baroque Angels – because their bulbous and flowing lines reminded people of the carved wooden figures in south German and Austrian churches of the Baroque period (seventeenth and early eighteenth centuries).

The first 501s had an updated version of the pre-war 2-litre six-cylinder engine. In 1954, this was supplemented by a 2.6-litre V8 with around 50 per cent more power; the six-cylinder 501 was also uprated, and coupé and cabriolet bodies were announced.

In 1955, BMW announced the 502 range, basically the same cars with extra equipment. These had either an uprated 2.6-litre V8 or a new big-bore 3.2-litre V8. The 501 V8 remained unchanged, but the six-cylinder 501s were given an enlarged 2.1-litre engine. Then the 3.2-litre V8 was in turn uprated for 1957's 502 3.2-litre Super.

In 1958, the 502 models were renamed as the BMW 2.6-litre, BMW 3.2-litre and BMW 3.2-litre Super. 501 V8 production ended, but the six-cylinder 501s remained available with their original names.

Front disc brakes and the (formerly optional) servo became standard on the 3.2 Super in 1959 and on the ordinary 3.2 a year later. Then in 1961, the six-cylinder 501s ceased production. The 2.6-litre was renamed a 2600, given the servo and front discs, and joined by a 2600L with more power and better trim. The 3.2 became a 3200L and the 3.2 Super became a 3200 Super (also known as a 3200S). Its 160bhp V8 made it the fastest saloon then made in Germany and among the fastest in the world.

The last of 21,807 Baroque Angels were built in March 1964.

Pedigree: BMW Before the 5 Series

When BMW needed a volume seller in the 1950s, it turned to the Isetta bubble-car. The model certainly sold well enough, but it presented a strange contrast with the big V8-engined cars at the other end of the range.

In the first half of the 1950s, motor-cycles brought in large profits for BMW. This is an R50 model, dating from the middle of the decade.

Pedigree: BMW Before the 5 Series

The 600

The 600 was a logical progression from the Isetta which must have seemed like a good idea at the time, but sales of just over 34,000 in two years never really matched BMW's expectations. Part of the problem was the price – the 600 was only barely cheaper than the entry-level VW Beetle. But it was also undeniable that buyers in the late 1950s wanted cars that looked like cars, and were losing interest in economy models which suggested that their owners might not have much money. Without the short-lived vogue for economy cars that followed the Suez crisis of 1956–57 and its accompanying petrol shortages, the 600 might have flopped badly.

Designed by Willy Black, the 600 was unashamedly intended as an enlarged Isetta with more power and a 'proper' four-wheel configuration. Its front end was pretty much unchanged from the Isetta's, but the wheelbase had been stretched to accommodate four seats, and a conventional rear axle had been added. This introduced to BMW the semi-trailing-arm independent suspension that would be seen on almost every new model for the next four decades.

The extra size and weight demanded a more powerful engine than the Isetta's, and so the 600 had yet another motorcycle powerplant – this time the 582cc twin from BMW's recently defunct R67. Top speed was 64mph (103kph).

The 700

The 700 was really the car that pulled BMW around in the late 1950s. Once again it was an upward progression in size from what had gone before – this time, the 600 chassis was stretched. By the time it entered production, however, the 700 had become BMW's first unitary-construction car.

The 700 was again masterminded by Willy Black, the man who had designed the 600 that it replaced. Black drew on the company's motorcycle technology once again, although this time he enlarged the twin-cylinder engine of the R67 motorcycle to get the power he needed for this larger car.

Styling was by the Italian Giovanni Michelotti, and its themes certainly echoed those of his Triumph Herald, an exact contemporary of the 700. His first sketch was for a slant-roof coupé, which appealed to BMW although they wanted more room in the passenger cabin. Michelotti therefore sketched up a saloon variant – never as pretty – and the Bavarians decided to build them both. The 700 Coupé entered production in August 1959 and the 700 Saloon joined it at the end of the year.

Even though the 700 was more expensive than a VW Beetle, its chic Italian styling brought in the buyers. Over the six years of its production, the car sold more than 188,000 examples and became BMW's best-selling car since 1945.

Engine power increased over those six years, and from 1961 there was an upmarket Luxus version. Then in 1962, the 700 was renamed the BMW LS. Among the most desirable of these small cars is the Baur-built cabriolet, but the most exciting was the limited-production 700RS, a competition roadster of which just 19 were built between 1961 and 1963.

With these cars, BMW was aiming at a fairly wealthy clientele who wanted large and powerful cars, and inevitably the cost of the cars limited their sales success. To counterbalance this and to achieve volume sales, the company decided to go ahead with a licence-built version of the Italian Isetta 'bubble-car' in 1955. It sold quite well, and went on to inspire the larger and improved BMW 600 four-wheeler, but its sales were not enough to offset the losses sustained when motorcycle sales nose-dived in the mid-1950s.

These disparate model-ranges made it difficult for BMW as a car maker to present any kind of real corporate image in the fifties. On the one hand, there were tiny economy cars, and on the other, there were large saloons and exotic sports and grand touring cars. In between there was nothing, and that made it hard for the public to make sense of BMW. The company did recognize that it needed a car range to fill that yawning gap, not least because the medium-sized saloon sector was a particularly lucrative one. However, attempts to design such a car came to naught because the company lacked the money to get it into production. So throughout the fifties, there was no direct equivalent of the 5 Series cars which would later go on to make the company's reputation.

CRISIS AND REBIRTH

BMW began to claw its way back into the mainstream of the car market with a delightful new small car called the 700 which was introduced in 1959. It sold well, but it arrived too late to stave off a crisis.

The 503

The 503 was the model with which BMW hoped to crack the American market in the mid-1950s. The baroque Angel saloons were not going to sell well in the USA, but BMW thought that an elegant grand tourer with the saloon's running gear and powerful V8 engine might.

The 503 used the saloon's perimeter-frame chassis as well as its running gear, and it had an all-alloy body designed by Albrecht Goertz. Goertz was a German who had worked with the Raymond Loewy design studio on Studebakers in the late 1940s and early 1950s, then became a naturalized American and set up his own design studio. He was persuaded to submit designs for the 503 (and the 507) by BMW's American importer, Max Hoffmann. Tempted by Hoffmann's offer to take a Goertz-designed 503 in quantity, BMW embraced the Goertz proposals and showed a prototype at the 1955 Frankfurt Show. Production began the following May.

Sadly, the 503's styling was flawed. The long bonnet hinted at power, but was spoiled by an ugly snub nose incorporating the traditional BMW grille. Electric windows were advanced for the time, and the power-operated hood on the cabriolets was a first for a German car. But the saloon gearbox, mounted remotely from the engine and operated by a woolly column change, did the 503 no favours. After September 1957, the gearbox was mounted conventionally so that a floor change could be fitted.

The 503 was always an expensive and exclusive car, competing with such exotics as the (more costly) Mercedes-Benz 300Sc and the Bentley Continental. Production averaged 100 or so a year, and it is likely that no two examples were exactly alike. Only a handful were delivered with right-hand drive. Despite their aesthetic shortcomings, the cars are very much sought-after today.

Pedigree: BMW Before the 5 Series

The Isetta developed into the 600, which hinted at what Fiat were doing rather more effectively with their 600 Multipla.

Salvation came with new funding and a new approach. This is the Neue Klasse 1500 of 1962.

Pedigree: BMW Before the 5 Series

The Neue Klasse floor-pan and running gear were used as the basis of a new grand touring coupé. This 2000CS clearly shows the curious front end styling of the model.

By the end of the year, BMW was in a bad way after losing 15 million Marks on a turnover of 150 million. A general meeting of shareholders was called during December, and there was strong support for a proposal from the bank which was BMW's chief creditor that the company should sell out. The most likely buyer was BMW's arch-rival, Mercedes-Benz.

However, the proposal was blocked by a substantial minority of shareholders, who voted for a counter-proposal to find another source of funding that would enable BMW to remain independent. No doubt a good deal of wheeling and dealing was done behind the scenes but, over the next two months, two businessmen who already owned a substantial proportion of BMW shares began to increase their holdings. By the autumn of 1960, some two-thirds of BMW shares belonged to the brothers Harald and Herbert Quandt, and BMW once again had funds.

There had been some work done on a medium-sized saloon during the 1950s, but the whole project was scrapped after the Quandt take-over and a completely new project was initiated, led by engineering chief Fritz Fiedler. Right from the beginning, it was always known as the Neue Klasse, and it was designed to plug the gap between the 700 and the Baroque Angel saloons, that would remain in production until 1964. The Neue Klasse was designed from the outset to be the car that BMW so desperately needed – and it proved to be right on target.

The 700 had pioneered unitary construction and semi-trailing arm rear suspension at BMW, and both features were retained for the new car. Chassis engineer Eberhard Wolff chose MacPherson strut front suspension (the car was one of the earliest to have it), with an anti-roll bar to give good handling. Styling and body engineering were entrusted to Wilhelm Hofmeister, and he established a distinctive 'BMW look' for the car, with a low waistline, large glass area, slim roof pillars, flat bonnet and boot and straight-through wing lines. On the engine side, Alex von Falkenhausen resurrected his 1958 sketches for a 1-litre overhead-camshaft four-cylinder originally intended for the 700 range, and developed these into an oversquare 1.5-litre alloy-head masterpiece with built-in stretchability.

Though the 1500 was shown in prototype form at Frankfurt in 1961, it was nowhere near ready for production. Sales didn't actually begin until a year later, and even then the car suffered from a number of teething troubles. However, it was the excellence of its basic design – with good handling, a gutsy engine and distinctive looks – that pulled it through. BMW followed the 1500 with a bored and stroked 1800 in 1964, distinguished externally by additional chrome trim, and with a twin-carburettor 1800TI that renewed the public's taste for BMWs with high performance. A 1600 next replaced the 1500, and from 1966 there was a 2-litre 2000 as well, with a variant of the original 1500 engine that had initially been prepared for the 2000 and 2000CS coupés of 1965, that used the Neue Klasse's floorpan.

In 1966, the Neue Klasse saloon was cunningly developed into a short-wheelbase two-door model known initially as the -2 and later as the 02, and with this model's introduction BMW pushed the whole Neue Klasse four-door saloon range further up-market. The 02s took over the

---

The 507

The 507 is probably the most widely recognised classic BMW of the 1950s. Like its great rival the Mercedes-Benz 300SL, it was inspired by the US importer Max Hoffmann, who told BMW he could sell a high-performance sports car in large quantities if the company could deliver.

In 1954, Ernst Loof designed and built a prototype on the 502 chassis with a 2.6-litre V8 engine. However, an alternative style put forward by Albrecht Goertz at Hoffmann's suggestion won the day. The Goertz style was for a curvaceous roadster with optional hard top. It was a shape which has worn incredibly well over the years, and examples of the 507 now change hands for very large sums of money.

The production cars had the 3.2-litre V8 in twin-carburettor form with 150bhp or, for the USA only, with 165bhp. Acceleration and top speed depended on which of the three optional axle ratios was chosen, but the performance of a 507 was broadly comparable with that of the contemporary Jaguar XK140. BMW claimed a 507 was capable of 136mph with the tallest 3.42:1 gearing, although 120mph was probably nearer the truth.

Yet this remarkable machine was never a strong seller. One problem was cost; another was BMW's inability to get production under way. Despite a Frankfurt 1955 announcement, the first cars were not delivered until the next year. By then, the Mercedes had become too well-entrenched as the definitive supercar, and the 300SL coupé's mutation into a roadster model in 1957 removed the 507's most obvious advantage. Lack of boot space in the first cars was also a major failing, and BMW was forced to introduce a smaller 'optional' fuel tank to free up more room.

Just 254 507s were sold between 1956 and 1959, all with left-hand drive. Some of the very last had disc brakes at the front instead of the all-drum system.

Pedigree: BMW Before the 5 Series

Bigger saloons arrived in 1968, initially with 2.5-litre and 2.8-litre six-cylinder engines which would go on to do service for many years in the 5 Series cars. Pictured is the final development of the range, a 1974 3.3L with 3.3-litre engine and long-wheelbase bodyshell.

1600 engine that then disappeared from the four-door range to leave the 1800 as the entry-level model. So it was that when the Neue Klasse finally retired in the early seventies, it had given BMW a position in the medium saloon market with engines of 1800 and 2-litre capacity.

The Neue Klasse saloons all had derivatives of the same basic design of four-cylinder engine, but BMW knew it needed to move back into the big saloon sector and that to do this it would need a larger-capacity engine. So a big six-cylinder was designed and put into production in 1968, when it appeared in a completely new bodyshell styled once again by Wilhelm Hofmeister. The cars were called the 2500 and 2800, and their engines were also put into a facelifted version of the CS coupés that became correspondingly more expensive. In due course, the big six BMW engine would be stretched first to 3 litres and then beyond.

These big saloons helped to clarify the position of the Neue Klasse in the new and more coherent BMW range hierarchy. At the bottom end, the smallest and cheapest cars were the 02s. Right at the top came the big saloons and coupés, and thus the Neue Klasse were neatly bracketed as the

Pedigree: BMW Before the 5 Series

The running-gear of the big saloons was also employed in reworked versions of the big coupés, which had a longer and much more attractive nose section than the 2000C/CS models on which they were based.

medium-sized, mid-range saloons. It was with this strategy – closely based on Mercedes-Benz practice – that BMW would enter the Seventies. There was some shuffling of market positions, but the replacement ranges with their new model designations made the hierarchy even clearer. There would be 3 Series models as the smallest saloons, 5 Series cars as the medium range, 6 Series coupés and 7-series big saloons – and that strategy has continued to the present day (although the 6 Series coupés were replaced by new flagship 8 Series cars in 1989).

Meanwhile, work had started on a replacement range for the Neue Klasse. This time, BMW's plan was to consolidate its position in the medium-sized saloon sector. It would also move up-market, with the 2-litre version now becoming the entry-level model, and in due course it would have to embrace six-cylinder engines as well and to challenge Mercedes-Benz's domination of the medium-saloon sector. This was not something that BMW could have contemplated ten years earlier, and it was the success of the Neue Klasse that had brought the company to that point.

2 First Generation: The E12 Models, 1972–81

By the end of the 1960s, the medium-sized Neue Klasse saloons were the oldest models in the BMW line-up. Since their introduction in 1961, the BMW range had been developed with new coupés, initially with four-cylinder engines and latterly with six-cylinder types in restyled bodies; and big six-cylinder saloons had been introduced to take over at the top of the range. The range had also been expanded downwards with the introduction of the two-door '02' derivatives of the Neue Klasse saloons in the middle of the decade. Although the Neue Klasse saloons were still selling well, they would certainly be due for replacement in the early years of the new decade, and BMW set about designing their replacements in or around 1969.

The company was in expansionist mood by this time. The success of the sixties model ranges had brought renewed prosperity, and the company was looking for ways to invest in its own future. One option might have been to expand by purchasing another car manufacturer, and in 1968 BMW certainly did put in an unsuccessful bid for the ailing Italian car maker Lancia, that was bought the following year by Fiat. However, by the end of 1969 it was clear that BMW's best option was to expand production at its existing factory sites in Germany. So a new building plan began, and each of the three years between 1970 and 1972 saw 200 million Deutschmarks of investment go into the BMW factories.

This expansion plan obviously had implications for the products. Demand for existing models was high, and BMW judged that it could grow even further. However, it was obvious that new products intended for introduction in the early seventies would have to be designed with a wider appeal than the ones they replaced in order to guarantee that growth of demand. As a result, the car which was to replace the Neue Klasse would not be a direct replacement for it but a model that would have a wider brief.

As a first stage, the model would be moved very noticeably up-market. Whereas the Neue Klasse of the late sixties consisted of 1.8-litre and 2-litre models, the 2-litre edition of its replacement would be the entry-level model. Smaller engines would now be confined to the two-door '02' range and to its eventual replacement. The Neue Klasse cars had also all been four-cylinder types, but the replacement medium-sized BMW would also have six-cylinder engines. In this way, it would be aligned much more closely with the standard-setting medium-sized Mercedes-Benz saloons, that had engines ranging from 2 litres up to a 2.8-litre six-cylinder. In addition, BMW would make a special effort to expand sales in the USA, where the marque had already attracted an enthusiastic following, and the new safety

First Generation: The E12 Models, 1972–81

The original E12 5-series was a smart-looking car, with just enough of the sporting saloon in its appearance to stand out from the other medium-sized saloons of its time. Styling was by Paul Bracq. This is one of the first 520 models, with four-cylinder carburettor engine carried over from the old 2000 saloon. Between 1972 and 1976, the E12s had a flat bonnet front, and the driver's door mirror was mounted on the door panel.

and emissions legislation planned for introduction in the early seventies would almost certainly make it necessary to develop a special variant of the new BMW saloon for the US market alone.

However, funds were not entirely unlimited. The enormous expansion of BMW's plants was consuming a huge amount of capital, and it was also going to cost a huge amount of money to design and develop the new medium-sized saloon. So right from the beginning, BMW planned to use existing drivetrain and 'chassis' components wherever possible, and to put the major investment into a new saloon bodyshell. This put the onus of responsibility onto stylist Paul Bracq, who had joined the company from Mercedes-Benz. He of course had to work closely with the body engineers, whose responsibility it was to turn his ideas into sheet metal.

Throughout the sixties, BMW had enjoyed a good working relationship with the Italian styling house of Bertone in

First Generation: The E12 Models, 1972–81

The sleek and well balanced styling of the E12 was instantly recognisable as coming from BMW, and established expectations which influenced the design of every subsequent 5 Series car.

The 528 arrived in 1975, just a year before the facelift. This is one of the first examples, with the original flat bonnet.

First Generation: The E12 Models, 1972–81

Turin. It had been Bertone who had designed the bodywork of the 3200CS coupé early in the decade, and the styling house had also acted as consultant for the styling of the Neue Klasse saloons. So it was no great surprise when BMW turned again to Bertone for a styling prototype in 1969. Most probably, the car was intended primarily to generate ideas, but it ended up having more than a passing influence on the shape of the new medium-sized saloon.

Meanwhile, the 'chassis' engineers were working on the new car's suspension. Existing BMW products had MacPherson struts at the front and semi-trailing arms at the rear with spring struts mounted on the hub carriers, and this layout had proved very successful. For the new saloon, however, which was given the project code of E12, the MacPherson struts were angled rearwards by 12 degrees, and the wheel travel was increased by 20mm (0.7in) all round to improve the ride comfort – an important consideration for the US market. The existing 2000 layout of disc brakes at the front with drum brakes at the rear was considered adequate for the less powerful versions of the new car, but an all-disc installation was also prepared to suit the higher-performance variants. The 2000's ZF worm-and-roller steering was also taken over without major change for the E12.

To some extent, the dimensions of the new car were dictated by the choice of engines planned for it. The fact that both four-cylinder and six-cylinder engines would have to be accommodated influenced

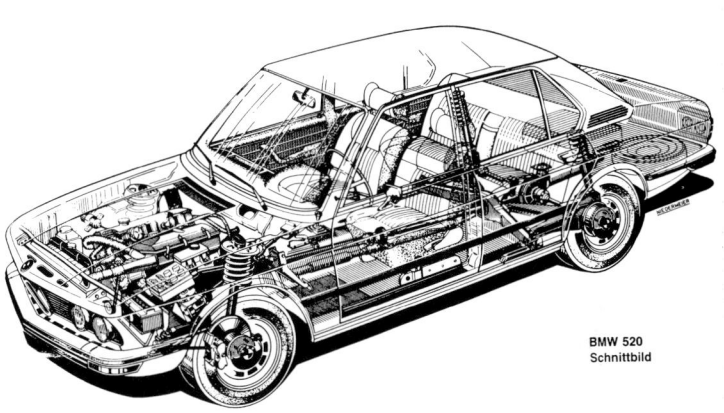

This BMW cutaway of an early 520 shows the overall layout of the car.

Bertone's Garmisch Prototype, 1969

Carrozzeria Bertone, SpA of Turin built a two-door concept car during 1969 on the basis of a left-hand drive BMW 2002ti. It was displayed at the Geneva Show in March 1970, where it bore the name of 'Garmisch' – a fashionable ski-resort in Austria. Although it had the 2002's short (1330mm, 52.4in) wheelbase, it was actually 4cm (1.57in) wider and 12cm (4.7in) longer than production examples of the 02, the extra length going into a larger boot. It was also 9cm (3.5 in) lower overall, which greatly reduced headroom.

Notable features of the Garmisch prototype were its large glass area and steeply raked front and rear screens. To protect the rear passengers from the sun, the rear window was covered by a curious honeycomb panel, which nevertheless allowed the driver to see the road behind. The interior included a futuristically styled dashboard and steering wheel, and a combined writing and make-up table pulled out of the passenger side of the facia.

First Generation: The E12 Models, 1972–81

Bertone's Garmisch prototype of 1969 was based on a 2002ti and was a two-door car. Elements of its styling were closely related to the production E12 shape introduced in 1972.

the length and shape of the bonnet, and to some extent the length of the wheelbase as well, which ended up some 75mm (3in) longer than that of the Neue Klasse. Overall length was dictated partly by aesthetic considerations, but also by the need to have a decent-sized boot, and was some 130mm (5in) up. Other dimensions were juggled to suit, and the E12 lost 25mm (1in) in its width and a similar amount in height as compared to the Neue Klasse. Even so, the end result was a much heavier car than the Neue Klasse, and the E12 in production form weighed between 110 and 135kg (roughly 250–300lb) more than the older car.

Around this package of dimensions, Paul Bracq concocted a masterful shape. It embodied the BMW family look, with a reverse-raked front end featuring the twin-kidney grille and twin headlamps, and it retained the clamshell bonnet that was a distinctive feature of the Neue Klasse and the 02 range. It also picked up the low beltline and large glass area that were established BMW features, while the general shape of the passenger cabin hinted at the big six-cylinder saloons and the rake of the front and rear screens recalled Bertone's two-door Garmisch prototype. The overall result was a svelte car with very clean lines – although these did date quite quickly and the shape had to be retouched towards the end of the decade.

The E12 bodyshell was not all about style, however. The importance of sales in the US market had persuaded BMW to develop a monocoque that would meet all the latest and foreseen crash-safety

First Generation: The E12 Models, 1972–81

regulations that were by this time proliferating across the Atlantic and, as a result, the E12 was designed with progressively deformable front and rear 'crumple zones' around a rigid passenger cell. This was not a new concept – Mercedes-Benz had patented the original scheme as long ago as 1951 and had been using it in production cars since 1953 – but it was a first for BMW.

Crash-safety considerations had also influenced the inside of the E12's passenger cabin, where the choice of deformable plastics and rounded edges reflected current thinking on these issues. Distinctive orange instrument illumination was also intended to contribute to safety at night. However, more immediately noticeable than either of these features was the striking new facia with its vertical central panel angled towards the driver. It incorporated the controls for a new and very effective heating and ventilating system, again carefully designed to provide the most desirable combination of cold air on the driver's face to keep him awake with hot air on his feet to keep him comfortable.

The individual front seats offered a range of adjustment fore-and-aft for the cushion and of rake for the backrest, and came with head restraints as standard. At the rear, meanwhile, there was a bench seat for three. Later on in E12 production, cushion height adjustment would be made available for the front seats along with individually shaped rear seats for two. For the moment, however, BMW planned to keep these refinements in reserve.

In fact, BMW's plan was to keep a whole variety of upgrades for the E12 range in reserve. At the car's introduction, only four-cylinder models would be made available. Once these had been accepted – and once new factory space had become available – the six-cylinder derivatives would follow.

Like many show specials, the Garmisch prototype was mainly an ideas car. The steeply-raked rear window had honeycomb slatting, which was intended to keep sun off the rear seat passengers' necks. Note the different wheels on this side!

First Generation: The E12 Models, 1972–81

Fortunately, BMW did not draw any inspiration from the Bertone car's deliberately futuristic interior.

The smallest of the six-cylinder engines would be the first to arrive, and the range would be gradually broadened and moved up-market as larger engines were introduced later on. By the end of the decade, BMW hoped to have a model-range that would match, model for model, the medium-sized Mercedes-Benz saloons – while the distinctively BMW blend of qualities would nevertheless give the cars a quite different appeal from those on offer from Mercedes.

The final stage was to give the new E12 range a distinctive public name. Marketing of the existing models had not been helped by the confusing array of model-names, and it was probably Marketing Director Paul Hahnemann who came up with the idea of clarifying the hierarchy by giving each of the BMW ranges a distinctive code number. Thus, the smallest saloons would become 3 Series cars, while the medium-sized models would become the 5 Series. Above these, the big six-cylinder saloons would become the 7 Series, and the prestigious coupés would fit into this scheme as the 6 Series. Within each series, the models would be further distinguished by their engine capacities, so that a 2-litre-engined 5 Series would be a 520 and a 2.5-litre-engined 5 Series would be a 525. These three-figure names had the further

First Generation: The E12 Models, 1972–81

advantage of recalling the much-respected BMWs of the thirties, with their names like 320, 326 and 328 (although those figures did not refer to engine capacity).

The decision was an important one. It brought order to the apparent chaos of the BMW model-ranges once the E12 5 Series had been followed during the seventies by the new 3 Series, 7 Series and 6 Series models. It also established a nomenclature which became the envy of other manufacturers, and during the eighties and nineties, several other companies attempted to emulate the BMW system.

THE 520 AND 520i, 1972

> The engine ... must be the smoothest and most fuss-free four-cylinder in the world ... But none of this extra-special refinement disguises the fact that the 520 really isn't quick enough for a £3000 car.
>
> *Motor*, 17 March 1973

> With fuel injection the car is not only much faster but infinitely more flexible and incredibly economical on fuel.
>
> *Autocar*, 28 June 1973

BMW's home town of Munich was to host the Olympic Games in 1972, and the company geared up to make maximum capital of the fact. It is well known that the distinctive towers of the new BMW administrative headquarters appeared again and again in TV pictures of the Olympic Stadium that had been built nearby and that commentators repeatedly explained what they were – thus providing BMW with free publicity worldwide. Less well known is that BMW also decided to launch its new 5 Series saloons at the Games, and that the worldwide media focus on Munich for the Games guaranteed the new cars maximum exposure. As it happened, the Games would later be overshadowed by the terrorist kidnap and

The facelift in 1976 for the 1977 model-year brought a taller grille and a revised shape for the bonnet's leading edge. There was also a bright edge to the grille on the six-cylinder cars, among which the 520 could now be counted. There were new mirrors, too, this time mounted to the leading edge of the window frame.

First Generation: The E12 Models, 1972–81

killing of a number of athletes – but by then, BMW had already reaped the publicity benefit it wanted.

Only two variants of the 5 Series were announced in 1972. The entry-level model was the 520, which had a 115bhp single-carburettor version of the 2-litre four-cylinder engine already familiar from the Neue Klasse 2000 and the 2002. Rather more expensive was the 520i, which had an injected version of the engine similar to that already available in the 2000tii and 2002tii. However, both engines had been developed beyond their Neue Klasse and 02 conditions. Both now had redesigned cylinder heads, with the tri-spherical swirl combustion chamber introduced earlier for the big six-cylinder types. The carburettor engine also had the latest type of low-emissions Stromberg carburettor with automatic choke, while the injected engine still employed Kugelfischer mechanical fuel injection, and would continue to do so until 1975.

The 520 offered customers the choice of a four-speed manual gearbox or a ZF three-speed automatic, the latter at extra cost. However, the 520i was conceived as a more sporting model, and only the four-speed manual was available. Despite its taller overall gearing, the 520i accelerated faster than the 520, reached a top speed some 6mph (9kph) higher, and yet drank no more fuel than the carburettor model. Those statistics were a powerful demonstration of the virtues of fuel injection, and it was obvious that this would be the way BMW would go in the future. Power-assisted steering was an extra-cost option on both cars, although it was probably specified on a higher proportion of injected models, purely because these appealed to more wealthy customers who tended to specify all the options when they ordered a car.

The new 5 Series saloons found immediate acceptance, although they were not introduced into all of BMW's markets at once. The USA, for example, would have to wait nearly three more years before getting its own version of the 5 Series BMW, and it would be two years before the South African production plant at Rosslyn turned out its first 5 Series cars. As BMW must have expected, the

---

BMW in the USA, 1975

Since their introduction to the USA during the 1950s, BMWs had been imported and distributed by Hoffmann Motors, the company owned by Max Hofmann who had influenced the development of cars like the fabulous 507 roadster. By the end of the 1960s, Hofmann Motors had successfully positioned the BMW marque in the USA as a desirable European import with its own distinctive image.

However, BMW's expansion plans for the 1970s envisaged a major growth of sales in the USA, and the company planned to keep all the profits from that growth in-house to fund further expansion and more new models. So in March 1975, the agreement with Hoffmann Motors was terminated and BMW North America was established as a branch of the German company. The new company had its headquarters in Montvale, New Jersey and established its own distribution network – a formidable task in a country as vast as the USA. It continued to market the existing BMW ranges, but among its first tasks was the marketing of the first 5 Series BMWs for the USA. These were the 530i models that were introduced just three months later in June 1975.

First Generation: The E12 Models, 1972–81

There was no mistaking the lines as those of a sporting saloon, even though five people could be carried in comfort. This is an early flat-bonnet car.

more expensive 520i proved to be the poorer seller, and the carburettor model accounted for some 80 per cent of sales over the next few years.

THE 525, 1973

> Power unit smooth and vigorous, but less eager than the 520i ... Very responsive fade-free all-disc brakes.
> *Autocar*, 17 August 1974

Meanwhile, building work at BMW's factories in Germany had continued apace. A second plant at Dingolfing (known as Plant 2-4) was ready for business by November 1973, and it was here that the company began production of its first six-cylinder 5 Series model.

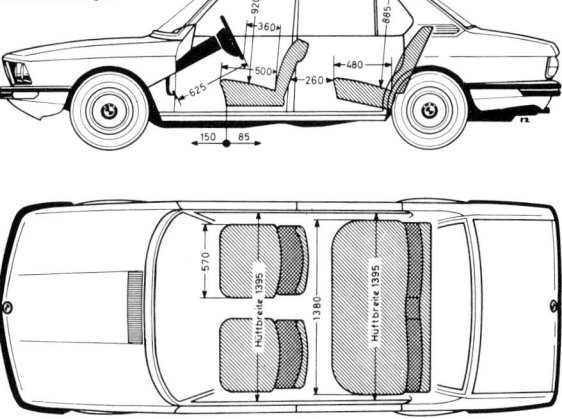

This drawing issued by BMW shows the interior dimensions of the original E12.

First Generation: The E12 Models, 1972–81

The 525 had actually been announced to the public at the Frankfurt Motor Show a couple of months earlier in September, and even then it had not been a great surprise; BMW had made no secret of the fact that a six-cylinder 5 Series car would join the 520 and 520i in due course. Like the four-cylinder cars, the new 525 came with an established and respected engine, in this case the twin-carburettor 2.5-litre six-cylinder from the big 2500 saloon. However, it was not quite the same engine; to maintain model differentials, it had a lower state of tune than in the larger and more expensive car. Only in 1976 would its 145bhp be raised to the 150bhp of the 2500, which by then had just 12 months of production life left.

The 525 brought more than the refinement and power of a six-cylinder engine. It also brought the height-adjustable front seats, individually-shaped rear seats, and all-disc braking system that had been held back at the initial launch of the 5 Series in 1972. Power-assisted steering was still not standard, being an extra-cost option exactly as on the four-cylinder cars, but the 525 did offer both four-speed manual and three-speed automatic transmission options. Overall gearing, at 3.9:1, was considerably taller than on the four-cylinder cars, and contributed to quite reasonable fuel economy. Even so, BMW had taken the precaution of installing a larger fuel tank with a capacity of 70 litres as against just 56 litres in the 520 and 520i.

That the 525 was exactly what BMW customers wanted was not in doubt when the returns for the first full year of production became available. Despite poor trading conditions for large-engined cars that had followed the fuel shortages consequent upon the Oil Crisis late in 1973, the 525 attracted 18,416 buyers as against 19,660 of the four-cylinder models during the 1974 calendar-year. Its success was enough to persuade BMW to press on with the next stage of the 5 Series development programme, which was to introduce an even more powerful version. In the meantime, however, the Oil Crisis had caused the company to take a step backwards with the 5 Series range, with the introduction of an entry-level 518 model at a price below that of the existing 520.

Alpina and the E12 5 Series

Alpina had been offering high-quality performance parts and performance conversions for BMW cars since the early 1960s, and it was no surprise that the company should turn its attention to the E12 5 Series cars. In the seventies, Alpina conversions were still quite rare, and perhaps for that very reason even more desirable than they are today.

There are two known Alpina conversions of the E12 5 Series cars. One involved fitting the 197bhp 3.3-litre engine from the big 3.3Li saloons into the E12 bodyshell, a conversion which was also carried out by Heidegger, the BMW importer and Alpina agent in Lichtenstein. The other, introduced in 1978 or thereabouts, was called the B7 Turbo and centred on a turbocharged version of the 3-litre six. This pumped out 300bhp at 6000rpm and 341lbft of torque at 2500rpm to give giant-killing performance of 0–60mph in 5.9 seconds and 0–100mph in 13.7 seconds. Naturally, the suspension and brakes were uprated to match this enormous power. Not many B7 Turbo conversions were made, and they are hugely desirable today even though their turbocharged engines do not give the seamless power delivery that is characteristic of later Alpina BMWs.

First Generation: The E12 Models, 1972–81

South African peculiarities: this is a 'facelift' 525, with the taller grille but without the door mirrors mounted on the window frame. Instead, there is a power-adjustable mirror mounted on the door panel!

THE 518, 1974

> Probably too sluggish for most keen drivers.
>
> *What Car?*, April 1982

It was in October 1973 when war broke out between Israel and an alliance of Arab nations. It was short-lived and victory went to the Israelis, but an immediate consequence was that the Arab oil-producing nations increased the price of oil by 70 per cent, cut back production and embargoed supplies to those nations they claimed had supported Israel during the conflict. Within weeks, there were major shortages of petrol worldwide, and sales of big-engined cars collapsed as buyers turned to smaller models that used less fuel. Diplomatic negotiations eased the supply situation early in 1974, but crude oil remained expensive and the price of petrol rose dramatically.

All of a sudden, fuel economy became a major item on the agenda for car makers. Faced with the very real possibility that big-engined cars would be unsaleable, many manufacturers scrabbled around in their parts bins to see what small-engined models they could come up with quickly

First Generation: The E12 Models, 1972–81

and cheaply. BMW's response was to fit the 90bhp 1.8-litre engine from the outgoing Neue Klasse 1800 into the 5 Series.

In the beginning, BMW had intended to confine engines of under 2-litre capacity to the 3 Series range that would make its bow in 1975. However, the public now wanted smaller engines, and BMW wanted to sell its 5 Series cars, so the 518 suited everybody when it was introduced in June 1974. In truth, the smaller engine did not offer any better fuel economy than the existing 2-litre, but the public perception was that smaller engines offered better economy and so the 518 sold on the strength of that, aided by suitably attractive pricing. During 1974 and 1975, the two years when sales of big cars were most affected by increases in the cost of petrol, the 518 sold surprisingly strongly, and although sales did dip as confidence returned to the market in 1976, the car sold far better than its mediocre performance really merited.

THE SOUTH AFRICAN 530

Meanwhile, 1974 also saw the introduction of the 5 Series models to South Africa, where they were assembled at the Rosslyn plant on the Cape out of kits shipped out from Germany. As far as the South African market was concerned, they were replacements for the BMW 1804 and 2004 models, that in turn were adaptations of Glas saloons dating back to 1964. BMW had bought the Glas concern in 1967, and with it the Dingolfing factory which now made 5 Series saloons, and had shipped the tooling for the Glas saloons out to South Africa. Equipped with engines and

This 520i, again built in South Africa, has the later style of steel wheels without nave plates.

First Generation: The E12 Models, 1972–81

This is the distinctive nose of an M535i, in this case lacking the Motorsport stripes which were often ordered.

transmissions supplied from Munich by BMW, these cars had established the BMW presence in South Africa, and during the seventies the company planned to build up this important market.

Rosslyn began with the assembly of 520 and 525 models, and began assembling 528s towards the end of 1974 in preparation for the car's 1975 introduction. Unaffected by the Oil Crisis and attuned to big-engined cars to suit long-distance travel in the vast African continent, the South Africans also decided to introduce their own special version of the 5 Series with the 3-litre six-cylinder engine from the big saloons and coupés.

So was born the 1974 BMW 530, a model made only in South Africa. Its 180bhp twin-carburettor engine gave it muscular performance beyond that available in any German-made BMW of the time, but there never seems to have been any intention of introducing a similar model in BMW's other markets. There was no need; the 2.8-litre versions of the 5 Series which BMW introduced in 1975 offered as much performance as the market wanted without the fuel consumption penalties of the larger engine. Nevertheless, the 3-litre 'big six' *did* appear in a German-built 5 Series, albeit with fuel injection instead of carburettors and offered for sale only in the USA.

First Generation: The E12 Models, 1972–81

THE 528 AND THE REVISED FOUR-CYLINDERS, 1975

> ... at its joyful best when allowed to rev ... the BMW 528 is the quietest and most refined car of this make that has yet been produced.
>
> *Autosport*, 21 August 1975

Oil Crisis or no Oil Crisis, BMW's customers still wanted the refinement and performance of the company's six-cylinder engines, and the success of the 525 gave the green light to the development of a 5 Series with the 2.8-litre carburettor engine first seen in the big saloons and coupés as long ago as 1968. To protect their sales, BMW initially offered it in slightly detuned form in the 528, where it had 165bhp instead of the 170bhp of the 2800, 2.8L and 2800CS.

The 528 made its bow in February 1975, just in time to gain star billing on the BMW stand at the Geneva Motor Show a month later. Four-speed manual and three-speed automatic options were available, and the overall gearing was higher than ever, at 3.64:1. Yet the 528 was also faster than ever, and the manual version could reach 60mph from rest in a whisker under 10 seconds and power on to a maximum speed of 123mph (198kph). This new-found performance was matched by the all-disc braking system already seen on the 525, and the 528 also pioneered low-profile tyres on the 5 Series, being equipped as standard with 195/70 HR 14s.

That year's other 5 Series revisions were held over until September and the Frankfurt Motor Show, however. For the new (1976) season, the four-cylinder cars were all given the 70-litre fuel tank from the six-cylinder models. The 518 engine gained a new carburettor and a slightly lower compression ratio, to the detriment of torque and fuel economy but not of power. The main reason for this apparently retrograde step appears to have been that demand had now built up for an automatic-transmission version, and that some changes in the torque characteristics were necessary to mate the engine satisfactorily to the three-speed ZF automatic already seen in the bigger-engined models. So from September 1975, a 518 automatic became available – and even though it lacked many of the driving characteristics traditionally associated with BMW, it sold well enough.

A drop in the engine compression ratio also affected the 520i, where the change accompanied a switch from Kugelfischer fuel injection to the latest Bosch K-Jetronic. Power and torque both went down, but the 0-60mph time was unaffected even though the maximum speed dropped from 114mph (184kph) to 112mph (181kph). BMW's main aim with these revisions had been to improve driving response in the mid-ranges where it mattered most in everyday use, and in this there was no doubt that they had succeeded. As for the 520, 525 and the still new 528, Frankfurt 1975 brought no important innovations.

SPECIALLY FOR THE USA: THE 530i, 1975

With the 525 and 528 models, BMW had begun to push the 5 Series range upwards into the territory occupied by the bigger six-cylinder saloons. In the USA, the cheapest of these was a model introduced in 1971 and known as the Bavaria, which was a slightly de-trimmed version of the 2800 saloon. And when the 5 Series was finally launched into the USA in June 1975, in the shape of a single model

First Generation: The E12 Models, 1972–81

South African models were different yet again. This M535i looks somehow even more purposeful than the bespoilered German-built car.

called the 530i, it was this model that it replaced.

A good deal of the basic architecture of the 5 Series had of course been designed to meet US crash-safety requirements. However, the US regulations were so far out of step with requirements in other markets that it was impossible to engineer the whole car purely to meet them. This was a problem which faced all car makers who sold their products in the USA in the early seventies, and it was not made any easier by annual changes in those regulations or by the fact that the State of California had stricter exhaust emissions requirements than the other 49 States and demanded a different state of engine tune. For all these reasons, it took BMW nearly three years to prepare the 5 Series for its US introduction.

There was no real problem in fitting the 5 Series with side marker lights to meet US regulations; it was simply a case of drilling extra holes in the wings and fitting red lights at the rear and huge amber lights (taken from the contemporary CS coupés) at the front. Nor was it too much trouble to fit extra warning lights inside, one to indicate that a seatbelt had not been fastened and the other to warn that the engine's emissions control gear needed a service. It was unfortunate that the lights could not have been integrated more neatly into the facia than they were, but the result was bearable.

However, making the car meet the low-speed impact regulations of the time demanded a lot more work. At this period, US law demanded that all new cars should

First Generation: The E12 Models, 1972–81

BMW E12 5 Series models, 1972–81

All models shared the same basic architecture of a unitary four/five-seater bodyshell with front and rear crumple zones, with a front-mounted engine driving the rear wheels.

518 (1974–81)

Engine
Cylinders	4
Bore and stroke	89mm x 71mm
Capacity	1766cc
Compression ratio	8.6:1
Carburettor	Solex 38 PDSI (Solex 32.32 DIDTA from Sept 75–Aug 80, Pierburg 2B4 from Sept 80)
Max. power	90bhp at 5500rpm
Max. torque	104lb ft at 3500rpm (103lb ft at 3,700 from Sept 75–Aug 80, 101lb ft at 4000rpm from Sept 80)

Transmission
(1974–75)
Gearbox	4-speed manual
Top	1.00:1
3rd	1.32:1
2nd	2.02:1
1st	3.764:1

(From Sept 75)
Gearbox	4-speed manual
Top	1.00:1
3rd	1.320:1
2nd	2.043:1
1st	3.764:1

(From Sept 75, optional)
Gearbox	3-speed ZF 3 HP 22 automatic
Top	1.00:1
2nd	1.52:1
1st	2.52:1

(From Sept 79, optional)
Gearbox	5-speed overdrive
Overdrive	0.813:1
4th	1.00:1
3rd	1.398:1
2nd	2.202:1
1st	3.822:1

Suspension and Steering
Suspension — Independent front suspension with MacPherson struts and anti-roll

bar; semi trailing arm rear suspension with coil-sprung stuts attached to the hub carriers and optional anti-roll bar

Suspension and Steering
Steering gear   ZF Gemmer worm and roller steering with optional servo assistance
Tyres   175 SR 14
Wheels   5.50J x 14

Brakes
Type   Servo-assisted with dual hydraulic circuit
Size   Front 280mm discs
                      Rear 250mm drums

Dimensions (in/mm)
Track   Front 55.3/1,406
                      Rear 55.6/1,412
Wheelbase   103.7/2,636
Overall length   181.9/4,620
Overall width   66.5/1,690
Overall height   56.1/1,425
Unladen weight   2,777–2,833lb (1,260–1,285kg) depending on specification

520 (1972–77) As for 518, except:

Engine:
Cylinders   1990cc four-cylinder
Bore and stroke   89mm x 80mm with overhead camshaft
Compression ratio   9:1
Carburettor   single Stromberg 175 CDET ;
Max power   115bhp at 5800rpm and 119 lb.ft at 3700rpm.
Transmission:
Gearbox   Four-speed manual gearbox
Top   3.764:1,
3rd   2.02:1
2nd   1.32:1
1st   1.00:1
Optional three-speed ZF 3 HP 22 automatic
                      2.52:1
                      1.52:1
                      1.00:1).
                      Axle ratio 4.11:1.

Suspension and Steering
Suspension   Rear anti-roll bar standard.
Rear track: 1,442mm (56.8in).
Unladen weight: 1275-1295kg (2810-2855 lb), depending on specification.

First Generation: The E12 Models, 1972–81

520 (1977–81) As for earlier 520, except:

Engine:
Cylinders	1990cc six-cylinder
Bore and stroke	80mm x 66mm with overhead camshaft
Compression ratio	9.2:1
Carburettor	Solex 4A1 carburettor
Max power	122bhp at 6000rpm
Max torque	118 lb.ft at 4000rpm.

Transmission:
Gearbox	Four-speed manual
Top	3.764:1
3rd	2.043:1
2nd	1.320:1
1st	1.00:1)

Optional three-speed ZF 3 HP 22 automatic
  2.56:1
  1.52:1
  1.00:1
  Axle ratio 3.90:1.

From September 1979:
Automatic ratios changed 2.48:1
  1.48:1
  1.00:1

Optional five-speed overdrive gearbox
  3.862:1
  2.002:1
  1.329:1
  1.000:1
  0.805:1

Five-speed close-ratio gearbox
  3.37:1
  2.16:1
  1.581:1
  1.241:1
  1.000:1

Suspension and Steering
Suspension	Rear anti-roll bar now optional only.

Unladen weight: 1350-1370kg (2976-3020 lb), depending on specification.

First Generation: The E12 Models, 1972–81

520i (1972–77) As for contemporary 520, except:

Engine:
Cylinders	1990cc four-cylinder
Bore and stroke	89mm x 80mm with overhead camshaft
Compression ratio	9.5:1
Carburettor	Kugelfischer PL 04 mechanical fuel injection
Max power	130bhp at 5800rpm
Max torque	131 lb.ft at 4500rpm (1972-1975).
September 1975 on	
Compression ratio	9.3:1
Carburettor	Bosch K-Jetronic fuel injection
Max power	125bhp at 5700rpm
Max torque	126 lb.ft at 4350rpm.

Transmission: No automatic option. Axle ratio 3.90:1.

Unladen weight: 1295kg (2855 lb).

525 (1973–81) As 520, except:

Engine:
Cylinders	2494cc six-cylinder
Bore and stroke	86mm x 71.6mm with overhead camshaft
Compression ratio	9:1
Carburettor	Zenith 32/40 INAT carburettors
Max power	145bhp at 6000rpm
Max torque	153 lb.ft at 4000rpm (1973-1976).
From August 1976	
Carburettor	single Solex 4A1 carburettor
Max power	150bhp at 5800rpm.

Transmission:
Gearbox	Four-speed manual
Top	3.855:1
3rd	2.202:1
2nd	1.401:1
1st	1.00:1)
Optional three-speed ZF 3 HP 22 automatic	
	2.56:1
	1.52:1
	1.00:1)
	Axle ratio 3.64:1.
From September 1979:	
Automatic ratios	2.48:1
	1.48:1
	1.00:1

First Generation: The E12 Models, 1972–81

Optional five-speed overdrive gearbox
    3.822:1,
    2.203:1
    1.398:1
    1.000:1
    0.812:1)
Optional five-speed close-ratio gearbox
    3.717:1
    2.403:1
    1.766:1
    1.263:1
    1.000:1

Brakes:
Size   Front 280mm discs
    Rear 272mm discs
Tyres: 175 HR 14 on 5.50J x 14 wheels
Rear track: 1,466mm (57.7in).
Unladen weight: 1380-1400kg (3042-3086 lb), depending on specification.

528 (1975–77) As for 520, except:

Engine:
Cyliners   2788cc six-cylinder
Bore and stroke   86mm x 80mm with overhead camshaft
Compression ratio   9:1
Carburettors   Two Zenith 35/40 INAT
Max power   165bhp at 5800rpm
Max torque   186 lb.ft at 4000rpm (1975-1976).
From August 1976:   Single Solex 4A1 carburettor
Max power   170bhp at 5800rpm.

Transmission
Gearbox   Four-speed manual
Top   3.855:1
3rd   2.202:1
2nd   1.401:1
1st   1.00:1
Optional three-speed ZF 3 HP 22 automatic
    2.56:1
    1.52:1
    1.00:1
    Axle ratio 3.64:1.

Suspension and Steering
Steering   Servo assistance as standard.
Tyres: 195/70 HR 14
Wheels   6J x 14

Brakes:
    Front 280mm discs
    Rear 272mm discs

Front track: 1,422mm (56in)
Rear track: 1,470mm (57.9in)
Unladen weight: 1415-1445kg (3119-3185 lb), depending on specification.

528i (1977–81) As for 528, except:

Engine:
Cylinders   Bosch L-Jetronic fuel injection
Max power   176bhp at 5800rpm
Max torque   173 lb.ft at 4300rpm (1977-1978).
From September 1978:
Compression ratio   9.3:1
Max power   184bhp at 5800rpm
Max torque   173 lb.ft at 4200rpm.
Transmission: From September 1978:
Automatic ratios   2.48:1
    1.48:1,
    1.00:1
Optional five-speed overdrive and five-speed close-ratio manual gearboxes
    (ratios as for 525 from 1979).
Axle ratio   3.64:1 with automatic
    3.45:1 with all manual gearboxes.
Tyres: From September 1978:
    195/70 VR 14.
Unladen weight: 1450-1485kg (3197-3274 lb), depending on specification.

530i (1975–78, USA ONLY) As for 528, except:

Engine:
Cylinders   2985cc six-cylinder
Bore and stroke   80mm x 88.9mm with overhead camshaft,
Compression ratio   8:1
Carburettor   Bosch L-Jetronic fuel injection
Max power   176bhp at 5500rpm
Max torque   185 lb.ft at 4500rpm.

Dimensions
Overall length: 4,823mm (189.9in).
Overall width: 1,706mm (67.2in);
Rear track: 1,460mm (57.5in)
Kerb weight: 1528kg (3368 lb).

First Generation: The E12 Models, 1972–81

M535i (1980–81) As for 528i, except:

Engine:
Cylinders	3453cc six-cylinder
Bore and stroke	93.4mm x 84mm with overhead camshaft
Compression ratio	9.3:1
Carburettor	Bosch L-Jetronic fuel injection
Max power	218bhp at 5200rpm
Max torque	228 lb.ft at 4000rpm.

Transmission:
Gearbox	Close-ratio five-speed manual (ratios as for 528i); no options.
Axle ratio	3.25:1 or 3.07:1.
Tyres:	195/70 VR 14
Wheels	6.50 J alloy

Dimensions
Unladen weight:	1465kg (3230 lb).

be able to withstand a 5mph impact at the front or rear without damage to the bodywork or lights, and the only way around this was to fit protruding bumpers which did nothing for the appearance of any car. The other major problem was making the engine meet the latest exhaust emissions regulations, and BMW already knew that the state-of-the-art emissions control gear of the day robbed engines of enormous quantities of power.

It was for that reason that Munich's engineers decided to fit the US-specification 5 Series with the 3-litre six-cylinder engine that was already available in that country in the bigger saloons. Fuel injection was chosen in preference to carburettors because it allowed more precise metering of fuel, which made for better control of exhaust emissions and optimum fuel efficiency. Further control of emissions was achieved by means of an Exhaust Gas Recirculation system, air injection and a thermal reactor.

This latter allowed the car to run on leaded fuel, which required a very rich mixture and a hot exhaust, achieved by retarding the spark at low speeds. So the 2985cc engine had its compression lowered to 8.1:1 from the 9.5:1 of the bigger saloons to avoid detonation. The result was an engine which put out 176bhp at 5500rpm and 185lb/ft of torque at 4500rpm. This was quite an achievement: although the European version of the engine without emissions control offered 200bhp and 200 lb/ft of torque, the 'detoxed' American engine still provided plenty of performance without the temperamental behaviour of some emissions-controlled engines from other manufacturers.

The Americans took to the 530i straight away, and bought 27,073 examples of the car between 1975 and 1978. Initial acceleration was very respectable, as the car took around 10.2 seconds to reach 60mph from rest. The 115mph top speed also matched customer expectations, although they did have to pay heavily for the privilege of owning this new BMW, that was more expensive than a home-grown Cadillac Sedan de Ville. Fuel economy of

First Generation: The E12 Models, 1972–81

Right at the end of E12 production, the South Africans introduced a 518i with the 105PS injected engine from the 318i. Note also the E28-style alloy wheels, another peculiarity of these final South African-built E12s.

the first cars was a sore point, too, and BMW quickly developed a more economical version. However, this did not meet the stricter Californian emissions regulations, and so West Coast buyers were still offered the original car while the revised model was sold in the other 49 States.

It would of course have been possible in theory for BMW to make a 530i for Europe by fitting the full-house 200bhp injected engine without the emissions-control equipment which robbed the US-specification car of 24bhp and 15lb/ft of torque. After all, the engine shared its block with the detoxed variant built for the USA, and there was therefore no question that it could be made to fit. In July 1977, the German magazine *Auto Zeitung* road-tested just such a car – a 525 fitted with the injected 3-litre engine. The car's straight-line performance was electrifying, but the magazine's testers did conclude that the chassis could not easily cope with all that torque on corners. Rear end breakaway occurred violently and without warning, and the overall verdict was that this was a car for the skilled driver as well as the connoisseur. Perhaps such considerations had also affected BMW's decision not to offer a version of the 5 Series with that much power – at least for the time being.

45

First Generation: The E12 Models, 1972–81

A large boot emphasised the family practicality of the E12. Note the characteristic BMW kit of small tools, carried in a special compartment under the boot lid.

First Generation: The E12 Models, 1972–81

MID-LIFE FACELIFT, 1976

With a range now spanning engine capacities from 1.8 litres to 2.8 litres in Europe, it was more than ever obvious that BMW's 5 Series was lining up as a challenger to the established medium-sized Mercedes-Benz saloons. And that became even more obvious in August 1976 when BMW announced a facelift for the range to help it retain its competitive edge against the latest mid-range Mercedes introduced at the Geneva Show in March that year.

The new Mercedes had been expected for some time, as the four-cylinder W115 and six-cylinder W114 saloons announced in 1967 were coming up for replacement. So BMW had been preparing a mid-life facelift for the 5 Series saloons for some time. It may be that some last-minute improvements were incorporated as a result of what Mercedes showed at Geneva, but the major elements of the August 1976 revisions had been planned well in advance.

Most obvious of the revisions was the new front end that took on more character through a raised twin-kidney grille and a bonnet with a raised centre section to match. New wheels with recessed centre sections lent the cars a more sporting appearance, and there were larger tail-light clusters. The fuel filler, positioned centrally in the rear panel on the first cars, was also relocated in the right-hand rear wing – the result, probably, of changes in the US crash-safety regulations. Lastly, the 1977 models had chromed door mirrors at the leading edges of the window frames (although the passenger mirror was an optional extra on most models) instead of the add-on mirrors that had always looked like an afterthought. These new mirrors had almost certainly been inspired by those which Mercedes had added to its medium-sized saloons in 1974. On some models, they were electrically adjustable. In South Africa, however, the locally-assembled cars had a door-mounted mirror for a time, without electrical adjustment.

There were also drivetrain revisions to the six-cylinder models of the 5 Series range in August 1976. Both the 525 and the 528 switched from twin Zenith carburettor to a single four-choke Solex, the 2.5-litre engine's power going up from 145bhp to 150bhp while the 2.8-litre went up from 165bhp to 170bhp. One reason for the change was undoubtedly easier servicing, as the Solex instrument did not need to be tuned and balanced in the same way as the twin Zeniths. However, BMW probably also hoped that the extra power would help to tempt a few more customers away from Mercedes' new 129bhp 2.5-litre 250, 156bhp 2.8-litre carburettor 280 and 177bhp 2.8-litre injected 280E.

THE SIX-CYLINDER 520 AND THE 528i, 1977

(The 528i is) ... a roomy yet reasonably compact family-size car of outstanding performance and with road manners which now have little cause for complaint to spoil the overall outstanding impression.

Autocar, 15 April 1978

That BMW was now throwing down the gauntlet to challenge Mercedes' domination of the medium-sized saloon market in Germany became even clearer at the next Frankfurt Motor Show, in September 1977. The four-cylinder 520 and 520i models were both swept away by a new 520 which featured a carburetted six-cylinder engine – one up on Mercedes

First Generation: The E12 Models, 1972–81

whose 2-litre engines still had only four cylinders.

It was quite clear that BMW had aimed for refinement with this new engine, known as the M-60, as it offered less power than the four-cylinder in the superseded 520i and had poorer straight-line acceleration. In South Africa, the change was not appreciated, and the six-cylinder 520 never did go into production over there; instead, the four-cylinder model remained available. Elsewhere, however, the more refined model was welcomed, and sales figures make clear that it quickly found more customers every year than the four-cylinder models had tempted.

Meanwhile, at the other end of the 5 Series range, BMW replaced the 528 with the 528i, that was essentially the same car but with the latest Bosch L-Jetronic fuel injection in place of its single Solex carburettor. Power went up to 176bhp from 170bhp while torque went down slightly and fuel consumption remained unchanged. But the 528i offered real gains in flexibility and in performance, bringing the 0-60mph acceleration time down by half a second to just under 9.5 seconds and putting the maximum speed up by just over 6mph to 129mph (208 kph). For a comfortable five-seater family saloon in the middle seventies, this was astounding performance, and the 528i quickly became a much sought-after car.

THE REVISED 528i AND THE US-MODEL 528i, 1978

Even so, the 528i package was not perfect, and BMW quickly responded to customer feedback. As a high-performance saloon, the buyers argued, what it needed was a geartrain better suited to its pretensions. So from the autumn of 1978, the standard four-speed gearbox in the 528i was supplemented by two five-speed options, one with an overdrive top gear and the other a close-ratio type in which fifth gear was the direct top. This latter box was made by Getrag and featured a dog-leg first gear. All three manual transmission options were now linked to taller overall gearing to give more relaxed high-speed cruising. At the same time, the ratios of the three-speed ZF automatic were changed to give buyers of that option a car with a more sporting feel. As for the engine, a raised compression ratio produced another 8bhp and moved the maximum torque slightly lower down the rev band. Maximum speed went up to 130mph and a further half second was lopped off the 0–60mph acceleration time.

It was also in 1978 that the 528i was introduced to the USA as a replacement for the 530i. More sophisticated emissions control equipment developed in the three years since that car had been announced now allowed BMW to offer the US market a 5 Series which was faster and more economical than before, even though its maximum power output was some 7bhp less. With a three-way catalytic converter and a Bosch lambda sensor in the exhaust instead of Exhaust Gas Recirculation, a thermal reactor and an air pump, the 2.8–litre engine was also very much more tractable than the earlier 3-litre type, and the same version could be sold in all 50 States. No matter: US customers still complained bitterly that they were being offered a car with a smaller engine, and in a country where big was still equated with beautiful, that was not acceptable!

Nonetheless, US buyers got over their initial reactions to the new model when they got behind the wheel. Sadly, BMW had still not found a way to overcome the need for those hideous 5mph bumpers

First Generation: The E12 Models, 1972–81

Introduced in 1978, the US-model 528i disappointed US enthusiasts who did not look beyond the boot lid badge to discover more tractability and more power than the superseded 530i had boasted. Note the additional marker lights and impact-absorbing bumpers.

introduced on the 530i, but as so many cars sold in the USA during this period were similarly disfigured for the same reason, the 528i still qualified as a good-looking car as well as one that performed well, handled well, and offered very high standards of comfort and build quality. In the four years during which the car was available in the USA, manual-transmission versions outsold automatics by a margin of around 10 per cent, confirming that the car appealed both to the traditional comfort-seeking US buyer and to the enthusiastic driver who had bought the car for its driving qualities. It was this car, probably more than any other, that established BMW as a force to be reckoned with in the compact saloon sector of the market across the Atlantic.

VERY HIGH PERFORMANCE: THE M535i, 1979

... a car of very high performance, with impeccable handling and roadholding. In spite of its fierce acceleration, its fuel economy is remarkable.

Autosport, 5 February 1981

First Generation: The E12 Models, 1972–81

The impact-absorbing front bumper and large amber side marker lights of a US-model E12 are seen here in close-up.

Even though the 1978 M1 supercar project had turned into something of a fiasco, careful management had allowed it to gather maximum publicity benefit for BMW. One result was that BMW Motorsport, whose M designation was found in the car's name, found itself riding high on public esteem.

So it was that the company looked around for ways of capitalizing on this. It was Jochen Neerspach, the former racing driver who headed the Motorsport division in the late seventies, who suggested that the M designation should be attached to a low-production, high-performance version of the 5 Series saloons. The sort of thing he had in mind was a production equivalent of Alpina's bespoke conversions, and the idea caught the imagination of the BMW management, who were acutely aware that the 5 Series shape was ageing quickly and that an attractive new high-performance variant might help keep sales alive until the replacement E28 model was ready in 1981.

It was not difficult to decide what to do. BMW's largest and most powerful engine in the late seventies was the 3.5-litre six-cylinder used in the 635CSi coupé and slated for 1979 release in the 735i saloon. The engine had a sporting pedigree, too, having been used in further developed four-valve form in the 635CSL racing coupés and in the M1 itself. The Motorsport division however saw no need to use the 277bhp four-valve version of the engine in the 5 Series body; adequate performance at a considerably lower price would be available with the 'cooking' 218bhp two-valve version.

The car known as the M535i was announced at the Frankfurt Motor Show in September 1979 as the first co-operative venture between BMW's production and Motorsport divisions. Engines and bodyshells were supplied by the production people to the Motorsport engineers, where they were assembled into the fastest 5 Series yet, capable of 136mph (215kph) and 0-60mph in 7.5 seconds or less. Naturally, there was a lot more to the M535i than a simple assembly operation, however: the car had been properly developed by BMW Motorsport to make the best use of its new-found power.

Thus the M535i had ventilated front disc brakes with four-piston calipers in addition to the solid discs at the rear seen on the 525 and 528i models, while power-assisted steering and a limited-slip differential were standard. High-speed stability was assisted by an air dam under the front bumper and by a small spoiler on the trailing edge of the boot lid, while tyres were the same 195/70 x 14 size as worn by the 525 and 528i, but VR-rated on the M535i to cope with the car's higher speed potential. The gearbox was the close-ratio Getrag five-speed type introduced the previous year as an option for the 528i, and alternative rear axle ratios of 3.25:1 and 3.07:1 were available.

M535i drivers were also treated to a special interior, which had hip-hugging Recaro seats at the front and a leather-bound steering wheel rim. On the outside, there was no mistaking the car, either, because it was delivered as standard with strikingly attractive cross-spoke alloy wheels made by BBS/Mahle in Germany. The result was discreet, but there were customers who still remembered with fondness the garish warpaint that had been applied to the BMW 2002 turbo back in 1973, and for them BMW would apply stripes on the M535i's air dam, sides and boot lid in the red, blue and light blue of the Motorsport division.

The M535i was never made available in the USA, mainly because its engine had not been homologated (that is, approved after testing) for that country. The US homologation tests were expensive, and BMW probably reasoned that sales would never justify the cost of preparing the car for the USA. So the M535i was sold in Europe only, and just 1,410 were built between the start of production in April 1980 and its end in May the following year. Among them were a number of right-hand drive examples for Britain, where BMW had a strong following and sales justified the production of what was really a very low-volume production variant. The main drawback of the right-hand drive cars was that the dog-leg first of the Getrag gearbox made for a difficult first-to-second change which could hinder rapid getaways; in left-

First Generation: The E12 Models, 1972–81

hand drive form, the change was rather easier and hardly occasioned comment.

Right-hand drive M535i models were also made in South Africa; introduced in 1981, they remained available for a time after the German-built car had gone out of production and bridged the gap to the new E28 models that were introduced in South Africa some time after their launch in other markets. They had a number of visual differences from the German-built models, however.

THE FINAL YEARS, 1979–81

> For families who didn't want a really big car, the 5 Series was an understated choice, filling much of the solid need that Rover once occupied in British hearts. There was, and continues to be, a sporting element among customers, but most are very conservative, right in the Mercedes mould.
>
> *Motor Sport*, August 1981

The South African-built cars continued to have their own peculiarities in this period, and in fact remained in production after the E12 had been replaced by the new E28 series in other markets. This is why the final examples had the E28 style of alloy wheels. South Africa also built a unique model in the 518i that was fitted with the 105PS four-cylinder engine introduced for the European 318i in 1980.

Not many changes were made to the mainstream E12 range during its final two years, however. Frankfurt in 1979 saw the five-speed overdrive and Getrag close-ratio five-speed gearboxes made optional on the 520 and 525 models, while the five-speed overdrive gearbox was made optional for the 518. The same show a year later introduced a further revised 518 with improved fuel economy from a higher compression ratio and a Pierburg carburettor in place of its earlier Solex. However, there was little point in making further changes: sales were slowing in the wake of the second Oil Crisis of 1979, but the brand-new E28 5 Series was just around the corner. When these second-generation BMW medium saloons were introduced in June 1981, the E12 models went out of production. What they had achieved was to give BMW a credible competitor to the class-leading Mercedes-Benz medium-sized saloons, and what BMW intended to do during the eighties was to press Mercedes even harder for the market leadership.

3 Second Generation: The E28 Models, 1981–87

The E12 range of 5 Series BMWs had been on sale for barely three years when work began on a revised and updated car. Encouraged by the success of its new model, the company was determined not to lose the advantage it had gained by allowing its rivals to catch up before a new model was ready.

Yet, paradoxically, when the new model arrived six years later in 1981, it was criticized for looking too much like the old one. No matter that much of its engineering was new, and no matter that there were many important changes under the skin; the fact was that it was not easy to distinguish the new E28 models from the old E12 types at a glance. That may be one reason why the E28s had a shorter production life than the E12s, lasting for only six years as compared to the nine of the earlier range.

Yet the E28 models took the 5 Series range into new territory. Unlike the previous range, it was planned with a four-cylinder 518 entry-level model right from the start. It also became the first BMW range to encompass diesel-powered models, and in due course it went on to include the astounding M5 – a family saloon with the performance and handling of a thoroughbred sports car. The diesels and the M5 also expanded the whole concept of the 5 Series range and allowed it to appeal to a wider audience, thus drawing in more and more customers and establishing the range even more firmly than before.

DESIGN AND DEVELOPMENT

The best way of understanding what went into the E28 cars is to look at the problems that faced the BMW design engineers under Karlheinz Radermacher when they started work on the new 5 Series in 1975. First, the medium-sized saloon market was traditionally rather conservative in its outlook and tended not to take to radical innovation. So BMW knew that it must not stray too far from the pattern established by the E12s.

Perhaps even more important was that the huge increase in petrol prices that had followed the Oil Crisis of 1973–74 had made all motor manufacturers focus on the question of fuel economy. By 1975, this was very much in the forefront of BMW thinking, but there were constraints on what the company could do about it. The existing production engines were relatively new and would have to have a long production life if they were to amortize their own development costs. A few tweaks could be made here and there to make them more fuel-efficient, but a further constraint was that BMW could not afford to sacrifice the performance on which its reputation depended so heavily in favour of better fuel economy.

Second Generation: The E28 Models, 1981–87

For the future, there could be new and more fuel-efficient engines that did not make performance sacrifices, but they were still a few years off. One of them would be a high-performance diesel engine – the company's first – on which work started in 1975. However, in the shorter term the engineers had to look for better fuel economy from improvements in other areas, and the most obvious one to tackle was weight. If the cars could be made lighter, the existing engines would not have to work so hard and would therefore deliver better fuel mileage. Lighter weight would also lead to better performance, which would be expected of the next generation of BMWs. As a result, a great deal of the E28 development effort went into weight-saving, and the production cars were 100kg lighter on average than the E12s that had preceded them.

So evolution, not revolution, was the keynote of the new E28 models. The door and roof pressings of the E12s were retained (which explains why the new cars looked so much like the old), but the wheelbase was shorter, the bonnet-line was lower and the boot-line higher and flatter. The bonnet itself was designed as a conventional crocodile type, mainly because this was inherently lighter than the clamshell type associated with BMWs since 1961. The higher line of the boot of course also made room for more luggage. The whole made for a more streamlined shape with lower aerodynamic drag, all of which made its own contribution to better fuel economy at high speeds.

BMW was convinced of the importance of slippery shapes, and had already started work on its own wind-tunnel. However, this was not ready in time to be used during E28 development, and so wind-tunnel tests were carried out elsewhere. They were the reason why the E28 range boasted a Cd of 0.382 to 0.384, some 12 per cent better than the average of 0.44 for the E12 range. They were also the reason why the E28

The E28 models looked a lot like the outgoing E12s in side view, but there were easy recognition points. Note the air intake on the D-pillar here (E12s had a black vent at the bottom of the pillar), and the indicator repeater on the front wing. This is a 535i.

Second Generation: The E28 Models, 1981–87

The neater and sleeker front profile featured a crocodile-type bonnet instead of the E12's clamshell, and the inner headlamps were smaller in diameter than the outer pair (except on US models).

went into production with a front lip spoiler, with flush-fitting trims on its standard wheels, and an undertray below the engine. All of these helped to manage the airflow around the car when it was moving, and they had the effect of reducing high-speed lift and making the E28s much more stable at speed than the E12s.

While redesigning the 5 Series bodyshell, BMW also took the opportunity to improve noise suppression and crash deformation. Analysis of load paths enabled the designers to make the front and rear crumple zones deform more progressively, and that same analysis allowed them to ensure that less noise was carried through the structure and into the passenger cabin. New double door seals also played their part in minimizing noise intrusion.

Some elements of the new car were borrowed from other BMW designs. Among them were the integrated door mirrors and the headlamps. These consisted of 7in main beam units outboard of 5.75in dipped-beam units, an installation prepared for the E23 7 Series range that would replace the existing big saloon models in 1977. A new double-link front suspension had also been designed for the E23 7 Series cars, and this was carried over for the E28s. Still fundamentally a MacPherson strut design, it differed from the conventional type because it was located not by a wishbone or a transverse link but by two separate links, ball-jointed together about two inches apart at the strut end. The advantage it gave was in the degree of offset, that both improved steering feel and left room for larger brakes if they were needed. However, for the 5 Series it was developed a stage further, and the fore-and-aft loads were taken by a rearward-facing lower link in order to minimize the potential damage from a front-end collision.

The basic design of the E12s' semi-trailing arm rear suspension was carried over, however, much to the disappointment of many commentators who felt that the system was too prone to sudden oversteer.

Second Generation: The E28 Models, 1981–87

Left: *There was no mistaking an E28 from behind with those big lamp clusters.* Right: *The first four models, introduced in 1981, could easily be distinguished one from another by their boot-lid badging.*

BMW's view was that the system was adequate for the performance of all but those derivatives with the highest performance. So for the new 528i (and for other, faster, models to follow), an anti-roll bar was fitted together with trailing arms angled at a steeper 13 degrees to the horizontal (the standard system had them at 20 degrees to the horizontal). There was also a trailing linkage that picked up where the arms met.

The familiar ZF worm-and-roller steering was kept in the specification for all the six-cylinder cars, mainly because BMW believed it gave a sharp enough response with their front suspension to make a rack-and-pinion system unnecessary. Power assistance was provided by Citroën-style high-pressure hydraulics, that also provided power for the brakes; an engine-driven pump kept the fluid under pressure in an accumulator, and release valves directed it to the appropriate system on demand. The rest of the braking system was also new: on the E12s, the hydraulic circuit had been split between front and rear wheels, but on the E28s it was split diagonally because this was better suited to the ABS system that BMW now planned to incorporate. The system would be optional on all the six-cylinder models except the 520i from the time of their launch in 1981.

The interiors of the new cars were redesigned, too, and BMW managed to make them more different from the E12 interiors than their use of the same passenger compartment would have suggested. As the 5 Series were family saloons, their rear seats were likely to be occupied a lot of the time, and so it was important that there should be plenty of room in the back. So a special effort was made in this area, and redesigning the seats brought an extra 27mm (1in) of rear headroom and an extra 40mm (1.6in) of rear leg room as compared to the E12 models. At the business end, meanwhile, the central section of the facia was now angled towards the driver, the steering wheel boss was smaller than before, and there was a less boxy instrument binnacle that blended more smoothly into the facia.

Electronics played a major part in the changes for the E28s. The E23 7 Series cars had introduced some sophisticated new electronic systems to the BMW range, and the plan now was to allow them to filter down gradually to the less expensive models. Broadly speaking, then, these new systems would be standard on the more expensive E28s and optional on the cheaper models. They included an on-board computer mounted in the facia. This allowed the driver to calculate such things

---

Engines in the E28 Models

The engines in the E28 models came from a variety of different BMW engine families.

Four-cylinders
The four-cylinder engines in the 518 and 518i were essentially the same engine, one with a carburettor fuel system and the other with fuel injection. Both displaced 1,766cc and both belonged to the M10 family that traced its origins right back to 1961 – although the 1.8-litre engine was not introduced until 1968. The M10 engines had a cast-iron block and an alloy head, with a duplex chain drive for the overhead camshaft. They were installed with a sideways tilt of 30 degrees.

Small sixes
The original small-six engine was the M60, developed specifically for the 3 Series and 5 Series cars in the mid-1970s. It was an oversquare design with a cast-iron block, alloy cylinder head, and a belt-driven single overhead camshaft. In the E28, it was installed with a sideways tilt of 20 degrees. The only M60 used in the E28 models was the 1,991cc engine in the 520i.

The M60 was developed during the 1980s into the M20, and an early version of this family was the 2,693cc 'eta' engine seen in the 525e and US-market 528e. However, the 'eta' engine had four main bearings instead of the seven of the other M20s (and M60s), in order to reduce internal friction. It was installed with a 30-degree sideways tilt, and was once again an iron-block, alloy-head engine with a single belt-driven overhead camshaft.

Big sixes
The big-block six-cylinder engines of the M30 family dated back to 1968. In the E28 range, they came in 2,494cc (525i), 2,788cc (528i) and 3,430cc (535i, M535i) forms. They had iron blocks and alloy heads, plus a single overhead camshaft driven by a duplex chain. They were installed with a 30-degree sideways tilt.

A derivative of the M30 engine was also found in the M5 (see Chapter 4). This used a bored-out M30 cylinder block but had a new alloy cylinder head with four valves per cylinder, activated by twin overhead camshafts. The camshafts were chain-driven, and the M88 engine was installed with a 30-degree sideways tilt.

Second Generation: The E28 Models, 1981–87

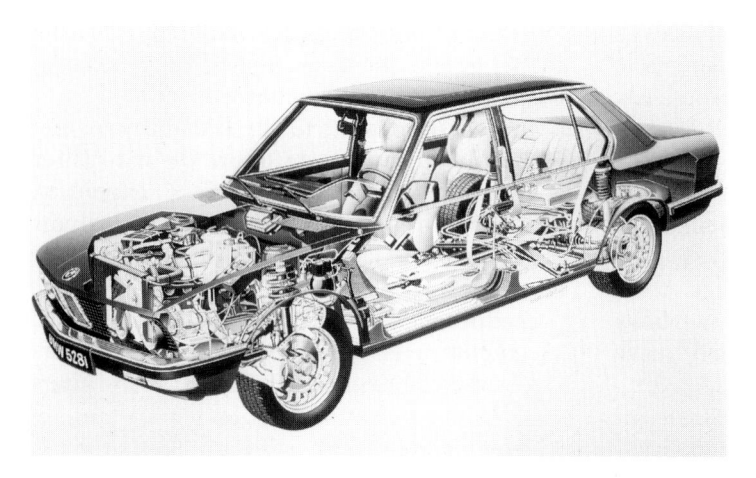

This cutaway of a 528i shows the main components of the E28, with its strut front and semi-trailing arm rear suspensions.

as distance covered, distance left to cover, fuel consumption for the trip, and average fuel consumption – and could also be programmed to act as a fairly sophisticated anti-theft device. Automatic temperature regulation was also installed, with a dial to select the required temperature. A fuel consumption indicator was also fitted to the dashboard, in the bottom segment of the rev counter, and gave an instantaneous read-out by monitoring the injector pulses in the constant-flow fuel injection system.

However, BMW decided to fit the dashboard of all the E28 models with the new Service Interval Indicator – a row of lights running from green through amber to red. This was the visible part of a 'black box' that gathered data from a number of points on the car and calculated when it needed service attention. The idea was that the red lights would remind the driver that a service was due, but would only light when the car really needed service attention. On a gently driven car, this might be at 9,000-mile intervals rather than the 6,000-mile intervals normally

recommended as a safety precaution for all cars – which would make for lower running costs. In addition, all models were planned to have an overhead console containing a battery of warning lights that monitored electrical circuits to indicate, for example, when a light bulb had failed.

THE 518, 520i, 525i AND 528i, 1981

By comparison (with the six-cylinder models), the 90bhp 518 seemed a somewhat fussy, underpowered and far less mechanically refined car.
Autocar, 20 June 1981
(The 520i has) one of the sweetest six-cylinder engines of all time ... (it is) good to drive, comfortable and beautifully built.
Motor, 28 March 1987
The smoothest and quietest model is the 525i ... though it feels restricted at the top end.
Road and Track, September 1981
In many ways, the new 528i is an exceptional car. Faster than its rivals and more economical, too, the car is the ultimate sports saloon ... but it lacks the ability to inspire absolute confidence in its driver – especially in the wet.
Autosport, 24 December 1981

The new E28 models were announced at the Munich Motor Show in June 1981. At first, they were available only with left-hand drive, and there were no special US Federal models. Right-hand drive cars became available in the autumn, and the first Federalized cars were ready early in 1982.

Second Generation: The E28 Models, 1981–87

There were four models in the initial E28 release, all of them using engines that had featured in the E12 range. The entry-level car was once again a 518, with the 90bhp 1.8-litre four-cylinder engine in the same state of tune as for the final E12s. However, this was the only four-cylinder car in the line-up, and the only model to have a carburettor-fed engine. The other three E28s had fuel-injected six-cylinder engines. The top-model 528i had the same 184bhp as in the E12 car of the same name, while that car's Lucas L-Jetronic electronic fuel injection with its over-run fuel shut-off had been added to the 2.5-litre engine to make the 525i. In this case, however, there was no increase in power, and the fuel injection simply made for greater flexibility and better fuel economy, the reason being that its 150bhp marked an important threshold in German insurance rates. As for the 520i, its engine was essentially the same M-60 2-litre type as in the final E12s, but with Bosch K-Jetronic mechanical fuel injection it now boasted 125bhp instead of 122bhp.

Four-speed manual gearboxes were still standard equipment except on the 528i, which came with a five-speed overdrive type. This was available at extra cost on the other E28s, and all the six-cylinder cars could also be bought with a three-speed automatic transmission. The five-speed overdrive gearbox proved a popular option on the 520i and became standard equipment on that model in September 1982, while the four-speed manual remained standard on the more expensive 525i. This apparently paradoxical situation probably came about because a large proportion of 525i buyers wanted the automatic transmission rather than a manual gearbox of any kind.

All four cars could be fitted with the 'sports suspension' option of gas dampers, stiffer springs and (on the 518) the stiffer

The high-performance image of the 535i persuaded tuning specialists to offer their own modifications. This car was done by TWR in Britain in the early 1980s.

59

Second Generation: The E28 Models, 1981–87

The BMW Diesel Story

In the mid-1970s, car manufacturers worldwide were still reeling from the shock of the first oil crisis and were desperately casting around for ways of improving their products' fuel economy. In common with many others, BMW saw diesel power as one solution, and it was in 1975 that engineering chief Karlheinz Radermacher asked his chief engine designer Karlheinz Lange to investigate diesel power.

Lange's brief in 1975 was an extraordinarily difficult one, and not only because of the lack of diesel knowledge and experience at BMW: despite an interesting flirtation with diesel power in the 1930s, BMW had never put a compression-ignition engine into production. Lange's biggest problem was that BMW had established a solid reputation for the excellence of its petrol engines, and that any diesel alternative that was not going to detract from the BMW image would have to be a lot more refined and offer a lot more performance than anything that was around in the mid-Seventies. However, Lange reasoned that if he started with one of BMW's refined petrol engines and adapted it for diesel operation, there was a reasonable chance that not too much of the base engine's refinement would be lost in the process. There would be other advantages, too: development time would inevitably be shorter than for a clean-sheet design, and the use of common components wherever possible would minimize the manufacturing costs.

So Lange chose as his starting point the latest M60 family of small-block OHC six-cylinder engines. These had first appeared in the early Seventies, and ranged in capacity from just under two litres to around 2.7 litres. What made them particularly attractive to Lange, apart from their high levels of refinement, was that there was enough metal between their cylinder bores to cope with the additional stresses imposed by the compression-ignition cycle.

Over the next three years, Lange and his team in Munich worked on dieselizing the M60 engine. No doubt rejecting direct injection as too unrefined, they designed a new cylinder head with indirect injection and Ricardo-type swirl chambers. These, they reasoned, would optimize combustion and therefore make for an engine that was not only fuel efficient but also less likely to fall foul of any future emissions control regulations. The missing ingredient now was performance, and the obvious way forward was to use a turbocharger.

While the BMW engineers were working on their new turbodiesel engine in the autumn of 1977, Mercedes-Benz announced their forthcoming 300SD turbodiesel model at the Frankfurt Motor Show. Not to be outdone, BMW showed its new engine to a select group of motoring journalists as soon as it had a viable prototype installed in a car. That was a year later, in October 1978; the car was an E12 5 Series saloon with five-speed gearbox, and it greatly impressed those who drove it. It reached a 112mph (180kph) top speed and took just 11.5 seconds to reach 62mph (100kph) from rest. Fuel consumption, according to a BMW spokesman at the time, was estimated at 20-25 per cent less than for a petrol car of comparable size and performance.

However, there was a problem. While the engine was almost production-ready, BMW had nowhere to build it. The company had embarked on a major expansion programme, and had simply run out of factory space. A solution seemed just around the corner when BMW formed a joint company with the Austrian Steyr-Daimler-Puch concern in March 1979 and started to build a new diesel engine factory at Steyr. However, there were delays and disagreements, and in March 1982 BMW bought out its partner's share in the company.

Some four years behind schedule, BMW now forged ahead with all speed to get its new turbodiesel six into production, and by March 1983 the first examples were making their way from Steyr to Munich to be fitted into the latest E28 5 Series cars. In the meantime, the production lines for the M60 petrol engines had also been moved into the Steyr factory. At a stroke, BMW had thus managed to fill a factory which was far too large for their diesel engine alone and to simplify their engine production lines – for the turbodiesel shared many common components with the M60 petrol six.

The production turbodiesel engine was known as an M21 and had a capacity of just under 2.5 litres. This size was chosen partly in order to avoid the punitive tax on bigger diesel engines that was levied in Italy – where diesel cars were becoming very popular and where BMW hoped to sell a large number of diesel-powered cars. Its 115bhp was remarkable for the time and equated to the 115bhp available from the 1990cc petrol six in the 520. In 1985, an 86bhp non-turbocharged version of the engine was put into production for the 3 Series saloons, and this later found its way into the 5 Series cars as well.

The original M21 turbodiesel engine was also sold to Ford in the USA, who needed an answer to Mercedes' success with diesel models in the luxury car market over there and offered it as an option in the Lincoln Continental and Mk.VII luxury barges. The BMW-engined Lincolns were introduced midway through the 1984 season but were withdrawn at the end of the 1985 season when the diesel bubble burst in the USA. Even though Ford did not therefore take the 191,000 engines originally planned, the 71,000 it did take made a big difference to the BMW coffers.

anti-roll bars of the six-cylinder cars. This normally came with the smart alloy wheels designed specifically for the new low-profile Michelin TRX metric-sized tyres. Steering was power-assisted as standard on all the six-cylinder cars, and the power assistance option could be had at extra cost on the four-cylinder 518. ABS was optional on the 525i and 528i from the beginning, but was not extended to the 520i until April 1982 and never would become an option on the 518.

Right from the start, the 520i proved to be the best-selling model worldwide of the new range, although of course the picture varied from country to country. The 528i lagged some way behind the 520i in worldwide sales, with the 518 next and the 525i the slowest seller.

THE 528e FOR NORTH AMERICA, 1982

There were ... two things we expect to find in a BMW that weren't in the 528e. One is the abundant engine smoothness that we've come to enjoy in six-cylinder BMWs ... The other ... is the lack of progress in the

The standard wheels on the first E28s carried bright steel centre caps. This overhead view of an early car shows the front wheel with its cap and – conveniently – the rear wheel without!

528e's styling. That's why the 5 Series is now one of the best in its class, while formerly it was the best.

Road and Track, February 1982

The North American market for BMW was very different from the market in Europe. In Europe, there were several different-engined versions of the 3 Series, 5 Series, 6 Series and 7 Series; in the USA, there was just one model available from each range, and that one was specially tailored for the

Second Generation: The E28 Models, 1981–87

The underbonnet view of a BMW 5 Series is always a pleasing sight. Note the finely-sculpted shape of the inlet tracts on the injected six-cylinder engine in an E28.

North American market to preserve the BMW image of a prestigious imported car with excellent performance and handling. So in the USA, there would never be and never had been a four-cylinder version of the 5 Series. Instead, US buyers were treated to just one 5 Series model – a six-cylinder developed specifically for the North American market – when the E28 was announced across the Atlantic in February 1982.

By this time the buying public's appetite had already been whetted by press reports on the European models. However, the car that North American buyers actually got seemed to present a different face of BMW from the one that everyone had been expecting. The 528e was tuned for fuel economy rather than performance, and there were howls of protest from buyers who felt they had been betrayed. Fortunately, BMW had something more exciting up its sleeve for the following year, but for a time a proportion of the US public certainly lost faith in the medium-sized BMW.

North American regulations had dictated a number of changes from the European-specification cars. First of all, the 5mph impact bumper regulations were still in force, and so the 528e wore extended bumpers. Fortunately, these had been much better integrated into the styling of the car than their equivalents on the E12 models for the USA. Side marker lights were only to be expected and were less obtrusive than before, and Federal regulations would not permit the smaller-diameter inner headlamps of the European cars, so the US models had four lamps of equal size.

Though its badging suggested that the 528e had a 2.8-litre engine under its bonnet, its engine capacity was actually just under 2.7 litres. Such had been the furore when BMW had switched from 530i to the 528i during E12 production, however, that the company's North American arm clearly decided it would be prudent not to invite trouble by drawing attention to the fact that the new engine was smaller yet again! That engine, in fact, never would appear in a car with badging that accurately reflected its capacity: when it was introduced in European models, the cars it powered were known by the 525e name!

It was that 'e' in the badging which was the key to the engine's purpose. BMW claimed that it was chosen to reflect the Greek letter eta, that is used in engineering to represent efficiency. However, it might just as well have been chosen to reflect the word 'economy', because that had been its design aim.

Second Generation: The E28 Models, 1981–87

The US variants of the E28 once again had impact-absorbing bumpers and side marker lights. These were much more elegantly integrated into the styling than they had been on the E12s.

Distinctive spoilers and TRX wheels instantly identified the M535i.

Under the US Corporate Average Fuel Economy (CAFE) regulations, a car maker had to pay a 'gas-guzzler' tax on their cars if the average fuel consumption of all the models it sold did not meet a certain standard. This meant that a small proportion of large-engined cars could exceed the consumption limit, but that they would have to be balanced by a large proportion of more economical smaller ones. As BMW wanted to continue selling its big 6 Series coupés and 7 Series saloons in the USA, the consumption of its smaller cars would have to balance out their thirst for fuel. So the 5 Series had to be made available with a notably fuel-efficient engine – and the result was the 528e.

The eta engine did not rev as high as the other BMW sixes, and was tuned to develop high torque at much lower crankshaft speeds. This reduced the frictional losses which cost fuel, and economy was further aided by the precise fuel metering now possible with the latest

Second Generation: The E28 Models, 1981–87

BMW E28 5 Series Models, 1981–87

All models shared the same basic architecture of a unitary four/five-seater bodyshell with front and rear crumple zones, with a front-mounted engine driving the rear wheels.

518 (1981–84)

Engine
Cylinders	4
Bore and stroke	89mm x 71mm
Capacity	1766cc
Compression ratio	9.5:1
Carburettor	Pierburg 2B4
Max. power	90bhp at 5500rpm
Max. torque	101lb ft at 4000rpm

Transmission
Gearbox	4-speed manual
Top	1.00:1
3rd	1.32:1
2nd	2.043:1
1st	3.764:1

(Optional)
Gearbox	5-speed overdrive
Overdrive	0.813:1
4th	1.00:1
3rd	1.398:1
2nd	2.202:1
1st	3.822:1

Suspension and Steering
Suspension	Independent front suspension with MacPherson struts and anti-roll bar; semi trailing arm rear suspension with coil-sprung stuts attached to the hub carriers
Steering gear	ZF Gemmer worm and roller steering with optional servo assistance
Tyres	175 SR 14
Wheels	5.50J x 14

Brakes
Type	Servo-assisted with dual hydraulic circuit
Size	Front 284mm discs
	Rear 250mm drums

Dimensions (in/mm)
Track	Front 56.3/1,430
	Rear 57.5/1,460
Wheelbase	103.3/2,625

Overall length	181.9/4,620
Overall width	66.9/1,700
Overall height	55.7/1,415
Unladen weight	2,557lb (1,160kg)

518i (1984–87)

As for 518, except:
Engine:
Compression ratio	10.0:1
Carburettor	Bosch L-Jetronic fuel injection
Max power	105bhp at 5800rpm
Max torque	105 lb/ft at 4500rpm.

Axle ratio: 4.10:1.

Dimensions
Tyres:	175 HR 14
Wheels	6J x 14
Unladen weight	1200kg (2645 lb).

520i (1981–87)

As for 518, except:
Engine:
Cylinders	1990cc six-cylinder
Bore and stroke	80mm x 86mm with overhead camshaft
Compression ratio	9.8:1
Carburettor	Bosch K-Jetronic mechanical fuel injection
Max power	125bhp at 5800rpm
Max torque	119 lb/ft at 4500rpm (1981-1985).
From September 1985	Bosch L-Jetronic electronic fuel injection
Max power	129bhp at 6000rpm
Max torque	126 lb/ft at 4000rpm.
From September 1986	Optional 'KAT' version with catalytic converter and Lambda sensor
Compression ratio	8.8:1
	Bosch Motronic electronic fuel injection
Max torque	118 lb/ft at 4300rpm.

Transmission:
Gearbox	Four-speed manual optional five-speed overdrive gearbox; optional ZF three-speed automatic
	2.48:1
	1.48:1
	1.00:1
From September 1982	Five-speed overdrive gearbox standardised and four-speed type no longer available

Second Generation: The E28 Models, 1981–87

From September 1983	ZF four-speed overdrive automatic replaced three-speed automatic option (ratios 2.73:1, 1.56:1, 1.00:1, 0.73:1).
From September 1985	Five-speed overdrive gearbox ratios changed
	3.72:1
	1.56:1
	1.00:1
	0.80:1
Axle ratio	3.91:1 (all models, 1981-1985, and 1985-1987 non-KAT automatic)
	4.10:1 (1985-1987 non-KAT five-speed)
	4.27:1 (all KAT models, 1985-1987).

Suspension and steering
Suspension:	Optional rear anti-roll bar.
Steering:	ZF (Kugelmutter) steering with standard servo assistance.
Tyres	175 HR 14
Wheels	5.50J x 14
From August 1985	
Tyres	195/70 HR 14
Wheels	6J x 14

Brakes:
From April 1982	ABS optional.
From February 1986	284mm discs at the rear.
From August 1986	ABS standardised.

Dimensions
Unladen weight:	1250-1320kg (2756-2910 lb), depending on specification.

524d (1986–87)

As for 518, except:
Engine:
Cylinders	2443cc six-cylinder
Bore and stroke	80mm x 81mm
Carbyrettor	Indirect-injection diesel with overhead camshaft and six-piston injection pump
Compression ratio	22.0:1
Max power	86bhp at 4600rpm
Max torque	110 lb/ft at 2500rpm.

Transmission:
Gearbox	Five-speed overdrive
Top	3.72:1
4th	2.02:1
3rd	1.32:1
2nd	1.00:1
1st	0.80:1
Axle ratio	3.91:1.

Suspension and steering
Suspension:	Rear anti-roll bar optional.
Steering:	ZF power-assisted (Kugelmutter).
Tyres:	195/70 HR
Wheels	6J x 14

Brakes: 284mm discs at the rear; ABS optional.

Dimensions
Unladen weight: 1335kg (2943 lb).

524td (1983–87)

As for 518, except:
Engine:
Cylinders	2443cc six-cylinder
Bore and stroke	80mm x 81mm
Carburettor	Indirect-injection turbocharged diesel with overhead camshaft
Compression ratio	22.0:1
	Garrett T 03 turbocharger and six-piston injection pump
Max power	115bhp at 4800rpm
Max torque	152 lb/ft at 2400rpm.

Transmission:
Gearbox	Five-speed overdrive
Top	3.72:1
4th	2.02:1
3rd	1.32:1
2nd	1.00:1
1st	0.80:1
Optional ZF four-sped overdrive automatic	
	2.73:1
	1.56:1
	1.00:1
	0.73:1
Axle ratio	3.15:1.

Suspension and steering
Suspension:	Rear anti-roll bar optional.
Steering:	ZF power-assisted (Kugelmutter).
Tyres:	
From August 1986	195/70 HR
Wheels	6J x 14

Brakes: ABS optional.
From February 1986   284mm discs at the rear.

Dimensions
Unladen weight: 1390-1410kg (3064-3108 lb), depending on specification.

525e (1984–87)

As for 518, except:
Engine
Cylinders   2693cc six-cylinder
Bore and stroke   84mm x 81mm with overhead camshaft
Compression ratio   8.5:1
Carburettor   Bosch Motronic electronic fuel injection; exhaust with catalytic converter and Lambda sensor
Max power   129bhp at 4800rpm
Max torque   166 lb/ft at 3200rpm.

Transmission:
Gearbox   Five-speed overdrive
Top   3.83:1
4th   2.20:1
3rd   1.40:1
2nd   1.000:1
1st   0.81:1)
Optional ZF four-speed overdrive automatic
                          2.48:1
                          1.48:1
                          1.00:1
                          0.73:1
Axle ratio   3.25:1 (manual)
                          3.46:1 (automatic).

Suspension and steering
Suspension   Rear anti-roll bar optional.
Steering   ZF (Kugelmutter) with servo assistance.
Tyres   195/70 HR 14
Wheels   6J x 14

Brakes   284mm discs at the rear; ABS optional.
Independent front suspension, with MacPherson struts and anti-roll bar; semi-trailing arm rear suspension with coil-sprung struts attached to the hub carriers.

Dimensions
Unladen weight: 1280-1300kg (2822-2866 lb), depending on specification.

525i (1981–87)

As for 518, except:
Engine
Cylinders	2494cc six-cylinder
Bore and stroke	86mm x 71.6mm with overhead camshaft
Compression ratio	9.6:1
Carburettor	Bosch L-Jetronic electronic fuel injection
Max power	150bhp at 5500rpm
Max torque	155 lb/ft at 4000rpm.

Transmission:
Gearbox	Four-speed manual
Top	3.855:1
3rd	2.202:1
2nd	1.402:1
1st	1.000:1

Optional five-speed overdrive gearbox (ratios as for 518)
Optional ZF three-speed automatic
	2.48:1
	1.48:1
	1.00:1
Axle ratio	3.45:1.
From September 1983	Five-speed overdrive gearbox standardised; four-speed type no longer available; ZF four-speed overdrive automatic replaced optional three-speed type
	2.48:1
	1.48:1
	1.00:1
	0.73:1
From September 1985	Axle ratio changed to 3.64:1 with five-speed overdrive gearbox.

Suspension and steering
Suspension	Rear anti-roll bar optional.
Steering	ZF (Kugelmutter) with servo assistance.
Tyres	175 SR 14
Wheels	5.50J x 14 steel wheels
or	195/70 VR 14
	6J x 14 alloy wheels.
From August 1986	200/60 VR 390
	165 TR 390 alloy wheels.

Brakes   284mm discs at the rear; ABS optional.

Dimensions
Unladen weight: 1800-1840kg (3968-4056 lb), depending on specification.

Second Generation: The E28 Models, 1981–87

528e (1982–83); USA ONLY

As for 525e, except:
Engine
Compression ratio	9:1
Carburettor	Bosch L-Jetronic electronic fuel injection
Max power	121bhp at 4250rpm
Max torque	170 lb/ft at 3250rpm.

Transmission
Gearbox	Five-speed overdrive gearbox only
Axle ratio	2.93:1.

Tyres	195/70 SR 14 tyres
Wheels	6.50J x 14

Dimensions
Overall length	4,800mm (189in)
Rear track	1,470mm (57.9in).
Kerb weight	1343kg (2960 lb).

528i (1981–87)

As for 518, except:
Engine
Cylinders	2788cc six-cylinder
Bore and stroke	86mm x 80mm with overhead camshaft
Compression ratio	9.3:1
Carburettor	Bosch L-Jetronic electronic fuel injection
Max power	184bhp at 5800rpm
Max torque	173 lb/ft at 4200rpm.

Transmission
Gearbox	Five-speed overdrive gearbox (ratios as for 518)
Optional ZF three-speed	automatic
	2.48:1
	1.48:1
	1.00:1
Axle ratio	3.25:1.
From September 1983	ZF four-speed overdrive automatic
	2.48:1
	1.48:1
	1.00:1
	0.73:1 replaced three-speed type.
Axle ratio	3.45:1 with this transmission only.

Suspension and steering
Suspension	Rear anti-roll bar standard.

Second Generation: The E28 Models, 1981–87

Steering	ZF (Kugelmutter) with servo assistance.
Tyres	175 SR 14
Wheels	5.50J x 14 steel
or	195/70 VR 14
	6J x 14 alloy wheels.
From August 1986	200/60 VR 390
	165 TR 390 alloy wheels.
Brakes	284mm discs at the rear; optional ABS.
From August 1986	ABS standard.

Dimensions
Unladen weight: 1830-1870kg (4034-4122 lb), depending on specification.

533i (1983–85); USA ONLY

As for 525e, except:
Engine
Cylinders	3210cc six-cylinder
Bore and stroke	89mm x 86mm with overhead camshaft
Compression ratio	8.8:1
Carburettor	Bosch L-Jetronic electronic fuel injection
Max power	181bhp at 6000rpm
Max torque	195 lb/ft at 4000rpm.

Transmission
Gearbox	Three-speed automatic optional
Axle ratio (manual)	2.63:1.
Tyres	200/60 VR 390 tyres
Wheels	65 TR 390 alloy

Dimensions
Kerb weight	1417kg (3125 lb).

535i AND M535i (1985–87)

As for 518, except:
Engine
Cylinders	3430cc six-cylinder
Bore and stroke	92mm x 86mm with overhead camshaft
Compression ratio	8.0:1
Carburettor	Bosch ME-Motronic electronic fuel injection exhaust with catalytic converter and Lambda sensor
Max power	218bhp at 6500rpm
Max torque	229 lb/ft at 4000rpm (without catalytic converter)
Max power	185bhp at 5400rpm

Second Generation: The E28 Models, 1981–87

Max torque	209 lb/ft at 4000rpm (with catalytic converter, non-US models)
Max power	182bhp at 5400rpm
Max torque	214 lb/ft at 4000rpm (US models).
Transmission	
Gearbox	Five-speed overdrive gearbox (ratios as 518)
Optional ZF four-speed overdrive automatic	
	2.48:1
	1.48:1
	1.00:1
	0.73:1 except for USA
Axle ratio	3.25:1 (manual)
	3.45:1 (automatic)
	2.63:1 (US models).
Suspension and steering	
Suspension	Rear anti-roll bar standard.
Steering	ZF worm and roller with servo assistance and quick ratio.
Tyres	220/55 VR 390 TRX
Wheels	165 TR 390 alloy
Brakes	Ventilated front discs
	284mm discs at the rear; ABS standard.
Dimensions	
Overall height	1,397mm (55in)
Rear track	1,470mm (57.6in)
Dimensions for US models as for 533i.	
Unladen weight	1450kg (3197 lb); US models' kerb weight 1465kg (3230 lb).

Bosch L-Jetronic fuel injection system. So despite the losses inevitable with a three-way exhaust catalyst, the 528e could travel for as far as 32 miles on a US gallon of petrol, and at its worst would always manage at least 20mpg – and the US gallon is some 8 per cent smaller than the Imperial gallon recognized elsewhere. However, the 528e was slower than the E12 528i that it replaced, with a maximum speed of just 111mph (as opposed to 122mph) and a 0-60mph time of 9.5 seconds (as opposed to 8.2 seconds). Arguably, this was of little consequence in a country with a blanket 55mph speed limit, but it mattered a great deal to Americans in terms of image.

Nevertheless, BMW had understood the American car-buying public better than the journalists writing for the specialist press. The 528e was aimed at those who wanted a BMW for its prestigious image and were not really concerned about its road performance. As BMW North America President John A. Cook described them at the 528e's launch, 'These people view themselves as creative, active, mobile and social. They are trend setters rather than followers. They project an image of being on the move rather than "I've made it".' And there were enough of them for the 528e to double 5 Series sales figures in the USA during 1982.

Second Generation: The E28 Models, 1981–87

Just visible on the grille of this UK-registered M 535i is the Motorsport logo. However, the car had no more performance than a standard 535i.

THE 533i FOR NORTH AMERICA, 1983

Wagnerian hot rod.
Car and Driver, February 1983

After the disappointment of the 528e in 1982, the arrival twelve months later of a 533i for the US market was a triumph. The car was basically a European-specification 528i with that car's rear suspension and ventilated front disc brakes, but with the bumper and lighting arrangements special to the USA and with a unique engine.

The 3210cc 'big six' was already available in the 6 Series coupés (633CSi) and 7 Series saloons (733i) and had thus already been homologated for the USA. Fitting it into the E28 bodyshell was thus a logical step, and in fact involved very little engineering complication. It made the 533i into the performance model which US buyers had expected in the first place, and was received across the Atlantic with a mixture of joy and relief. At 127mph (204kph), the top speed was even better than that of the late, lamented E12 530i, and the 0–60mph acceleration time of 7.7 seconds was once again in the league that US buyers had come to expect of BMW. A five-speed overdrive gearbox, rear anti-roll bar and low-profile tyres on alloy wheels made it quite clear that this was a driver's car in the true BMW tradition.

Even so, there was enough demand for an even faster 5 Series BMW in the USA for BAE Turboboost Systems of Torrance in California to consider the development of a turbocharged 533i worthwhile. The car was developed with the assistance of Suspension Techniques of El Monte in California, who improved the standard suspension to cope with the new-found performance from the turbocharged engine. With its Garrett AiResearch T3 turbocharger, the 3210cc engine delivered 250bhp at 5100rpm and 225 lb/ft of torque at 3900rpm. The company's prototype was tested by *Motor Trend* in November 1983, and achieved the astonishingly fast 0–60mph time of 7.72 seconds, but whether many cars were built for customers is simply not clear.

FOUR-SPEED AUTOMATICS, THE FIRST TURBODIESEL AND THE 525e, 1983

Whatever BMW may say about (the 524td) measuring up to the company's high performance standards, we think it misses

73

Second Generation: The E28 Models, 1981–87

The black rubber boot-lid spoiler of the M535i and the Motorsport logo made quite clear from behind what the car was.

the mark ... This car may exude the usual level of BMW competence, but where's the BMW fun?

Motor Trend, September 1985
With remarkable economy and no sacrifice in driving pleasure, BMW's capable executive saloon ... (the 525e) ... is the family man's answer to the Porsche 944.

Motor, 10 October 1983

The emphasis was on fuel economy for the 1984 5 Series models announced at the Frankfurt Show in September 1983. At that show, the range was supplemented by a diesel model called the 524td and by a new economy-oriented 525e, while a new automatic transmission for the six-cylinder petrol models brought an overdrive top gear and a lock-up for the torque converter to prevent fuel-wasting converter slip at cruising speeds.

This new transmission was once again made by ZF, and was known as the 4 HP 22, that first digit reflecting the fact that it

Second Generation: The E28 Models, 1981–87

In South Africa, however, tastes differed. The full bodykit was reserved for the M5 (see Chapter 4), while the M535i was distinguished from lesser models by its Motorsport alloy wheels and by the Motorsport logo on the grille.

The dashboard of a 524td model shows the Service Interval indicator which was new to the 5 Series with the E28 cars.

The centre console of the E28 dashboard was angled towards the driver. This is an M535i model (note the Motorsport colours on the lower steering wheel spoke.)

Second Generation: The E28 Models, 1981–87

The 524td was the first 5 Series diesel, and its turbocharged engine was based on the small-block petrol six.

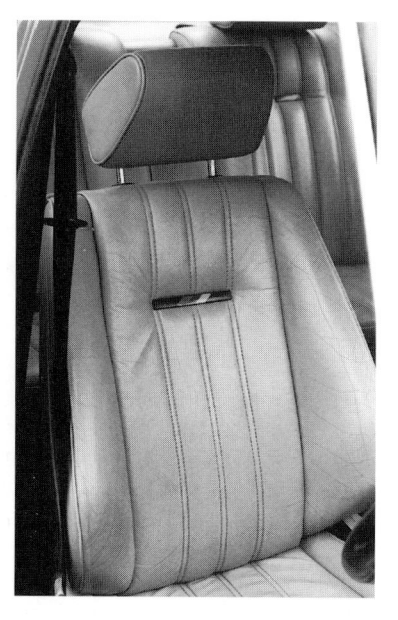

The Motorsport colours are displayed on this bar attached to the seats of an M535i.

had four speeds rather than the three of the older ZF 3 HP 22 transmission. Its overdrive fourth gear was engaged at road speeds of 85-90kph (53-56mph), although its engagement was delayed if the car was climbing a hill or accelerating hard. Change quality in the new gearbox was superb, and of course full manual over-ride was available. It replaced the older three-speed type as the optional transmission on the 520i, 525i, and 528i models, the latter taking on a lower axle ratio when equipped with the new automatic. At the same time, the 525i took on the five-speed overdrive manual gearbox as its standard transmission, thus removing an anomaly that had existed since the 520i had been given the five-speed as standard a year earlier.

The new diesel car had a turbocharged six-cylinder engine of 2443cc, and so was known as a 524td. The car was not made available in all of BMW's markets – there were no right-hand drive models for Britain, for example, where diesel cars were still viewed with suspicion – but those who were lucky enough to get the 524td recognized what a ground-breaking machine it was.

It was, of course, BMW's first production diesel car. Much more important, though, was the fact that the engine represented a major breakthrough in diesel engine design. Diesels had traditionally been frugal in their use of fuel, but the robust construction needed to withstand the stresses of the compression-ignition cycle had ensured that they were low-revving engines with slow responses to the accelerator pedal. Combustion noise had always been a problem, too, and for all these reasons diesels had been primarily associated with commercial vehicles or – as passenger car engines – with taxis.

With the new M-21 engine, however, BMW had changed all that. The engine was remarkably quiet, free-revving, and would respond quickly to the accelerator. It was no less durable than existing designs, and it also offered the excellent fuel economy traditionally associated with diesels. Its six cylinders ensured that it was as smooth as the customers expected from BMW, and its performance was quite remarkable. It gave the 524td acceleration broadly similar to that of the petrol-powered 520i. In no sense did it let down the BMW reputation for refinement and performance, and in countries such as Italy and France where diesel fuel was much cheaper than petrol, the car was a sure-fire winner. Only in the USA, where diesel cars were a short-lived fashion, did it receive a more muted welcome.

The M-21 engine also scored some points at the expense of rival Mercedes-Benz. The OM617A turbodiesel engine fitted to that company's 300D Turbodiesel and 300SD models may have pumped out 125bhp as against the BMW's 115bhp, but it was an inherently rougher five-cylinder design and it needed a full 3-litre capacity (actually 3,005cc) to achieve that, while the BMW engine had half a litre less swept volume at 2,443cc.

Like the turbodiesel Mercedes, the 524td was offered with automatic transmission – although BMW again scored a point by making the five-speed manual standard when Mercedes did not offer a manual transmission at all. There were no other serious competitors, however: automatic transmissions and diesel engines were not generally linked in cars at the time, mainly because the power losses through the automatic's torque converter combined with the low power output of the diesel engine to produce abysmal performance figures.

Sales of the new turbodiesel models never quite reached Mercedes-Benz levels for the 300D Turbodiesel at its peak, and this was no doubt partly because BMW was a newcomer on the diesel scene whereas Mercedes was well-established. Nevertheless, worldwide figures hovered around the 15,000 to 16,000 a year mark, with a best in 1985 of 22,667. There was no doubt, then, that the turbodiesel BMW was here to stay.

The second 5 Series debutante at the 1983 Frankfurt Show was the 525e, a model which was designed to bring excellent fuel economy to the 2.5-litre sector of the medium saloon market. It was really an improved and Europeanized version of the 528e that had been introduced for the US market some eighteen months earlier, and once again, its model-number did not reflect the 2.7-litre capacity of its 'eta' engine. This was slightly more powerful and slightly less torquey than the US engine, and came as standard with an exhaust catalyst in anticipation of new German regulations that would make tax exemptions for cars with 'clean' exhausts after 1 January 1986.

Most important was that the 525e introduced more new technology to the 5 Series range. This was the Bosch Motronic engine management system, available on certain 7 Series saloons since 1979. Motronic replaced the L-Jetronic injection system of the US-market 'eta' engine, but it was more than mere fuel injection. It was controlled by a small computer 'brain' which was also linked to the ignition system and could advance or retard the spark timing in order to ensure complete combustion under all conditions. Put simply, it guaranteed optimum combustion under a wide variety of operating conditions and therefore improved efficiency, leading to better power output and lower emissions levels. With a lower compression ratio than in the US engine, the European 'eta' six-cylinder developed slightly more power and slightly less torque than its US counterpart, and returned quite exceptional fuel consumption figures as well as offering more than acceptable performance. The British magazine *Motor* recorded nearly 112mph (180kph), 0–60mph (0–96kph) in 10.3 seconds and nearly 25mpg (11.3/100km) overall, which was a quite exceptional cluster of figures for the time.

HOW GERMANY PERSUADED THE CAR MAKERS TO DEVELOP CLEANER EXHAUSTS

The West German Government in Bonn had originally wanted to enforce the use of catalytic converters on all new cars from January 1986, but in 1984 it postponed the deadline until 1 January 1989 after protests from the German motor industry and from other EEC countries. However, cars with engines of 2 litres or larger had to comply from 1st January 1988.

Bonn advised German car makers to offer catalytic converters from July 1985, and from that date increased tax on leaded fuel while decreasing that on unleaded by a corresponding amount. This was one incentive to buyers; a second and more powerful one was that cars equipped with catalytic converters were excused from road tax up to 31st December 1988 (or 31st December 1987 for those with engines of 2 litres or larger). There was also an increase in road tax for cars without catalytic converters from 1st January 1986.

THE 518i, 535i, M535i AND M5, 1984

(The 518i is) certainly a big improvement on its carburetted predecssor, even if it's a little short of that old Bavarian magic.
Motor, 26 January 1985

(The 535i is) very nearly one of the great cars of our time for its combination of spacious accommodation, refinement and high performance. Only its handling, particularly in poor conditions, lets it down.
Fast Lane, March 1986

(The M535i is) a quite exciting formula for refined entertainment ... a unique recipe.
Autocar, 16 January 1985

The Frankfurt Show in 1984 was a major milestone for the E28 range, as BMW introduced no fewer than four new models, of which three took the range into new areas of high performance – and of cost. At the bottom end of the range, there was simply evolutionary change, as a 518i replaced the existing 518. However, at the top end of the range were two cars with the big-block 3.5-litre six-cylinder engine in two-valve form, and above them came a very special machine developed by BMW Motorsport and using that division's four-valve version of the 3.5-litre six. The M5 is dealt with in detail in the next Chapter.

Fuel injection had gradually been making its way down the BMW model hierarchy for several years. It had been standard on the US models since the 5 Series had been introduced there, mainly because it offered an effective way of controlling exhaust emissions. Outside the USA, however, it had mainly been used to enhance performance. On the E12 5 Series, it had been fitted only to the M535i and 528i models. With the arrival of the E28s, it had spread right down to

Second Generation: The E28 Models, 1981–87

the 2-litre 520i; and now BMW decided to fit it to the entry-level 1.8-litre car as well. So carburettors disappeared from the 5 Series range altogether.

There was still some sense of hierarchy within the 5 Series range, however. Motronic engine management was a more complicated and expensive system than the L-Jetronic electronic fuel injection, and so it was L-Jetronic that was fitted to the 518i. This brought it into line with the 525i and 528i models, which also had L-Jetronic injection, but left the 520i out on a limb with its older K-Jetronic mechanical injection. Adding the precise fuel metering of the L-Jetronic system to the 1.8-litre engine enabled BMW to raise its compression ratio as well, so that the injected engine gave a healthy 105bhp as against the 90bhp of the carburetted type in the 518. Peak torque was also slightly increased but was reached at higher crankshaft speeds, and the injected engine was generally higher-revving than the carburetted type. This different nature was matched by a raised axle ratio in the 518i.

The basis of the two 3.5-litre models was of course the 528i, and both of them shared its rear suspension, five-speed gearbox and optional four-speed automatic.

E28s on the assembly lines at the BMW (South Africa) assembly plant in Rosslyn.

Second Generation: The E28 Models, 1981–87

The selectable-mode automatic controls of an M535i.

Cockpit of an M535i: note the Motorsport colours on the lower spoke of the steering wheel, and the special seats.

Second Generation: The E28 Models, 1981–87

Both had ventilated disc brakes at the front to cope with their higher performance, both had ABS as standard, and both had fatter, lower-profile tyres on new alloy wheels. Both of them also had the same M30 3,430cc engine, that had first seen the light of day a couple of years earlier in the big 735i saloon. The 3.5-litre engine was a member of the same family as the 2.8-litre type in the 528i, and had been developed over the years through a series of size increments until it had both a larger bore and a longer stroke. The outside dimensions of the block were still the same, however, and so the engine fitted under the bonnet of the E28 with the minimum of adaptation.

However, BMW offered two versions of this engine, one equipped to meet the forthcoming German exhaust emissions regulations and one without the emissions-control gear. The emissions-controlled car carried a power-sapping catalytic converter in its exhaust system and was barely more powerful than a standard 528i, with 185bhp as against that car's 184bhp. Where it did score was in acceleration, thanks to 209 lb/ft of torque instead of 173 lb/ft. However, without its catalytic converter, the 3.5-litre engine revved up to 6,500rpm and put out a glorious 218bhp, while torque delivery was even better with 229lb/ft at 4,000rpm. In some countries – Britain, for example – the emissions-controlled engine was simply not available. In Germany, however, both versions were on sale.

BMW must have had some doubts about the sales potential of a car that offered a single extra brake horsepower from 700cc and cost a lot more than a 528i. The fact that the 3.5-litre engine gave much better acceleration would have been lost on a good number of potential buyers, as it has always been the case that the man in the street can relate to power but struggles with the concept of torque. So it was for this reason that the 3.5-litre engine (still in both cat-equipped and non-cat forms) was made available in two E28 models with very different characters. The ordinary 535i looked like any other 5 Series of the time, give or take its alloy wheels and fatter tyres. However, the M535i sported a full bodykit of spoilers and sill extensions to give it the image of a performance car. A breakdown of sales figures is not available, but it would be surprising indeed if many buyers of an M535i chose to strangle their cars' performance by specifying the lower-powered emissions-controlled engine. Most buyers with green sensibilities almost certainly went for the less overtly sporting 535i!

The M535i had been prepared by BMW's Motorsport division, although the main effect of that division's work was to give the car a more sporting character; it certainly had nothing like the performance of the M5, that had been given the full Motorsport treatment. What the M535i did have was slightly uprated springs matched to gas dampers in order to give it sharper handling. It also had front seats with adjustable bolsters, broadly similar to those in the high-performance 6 Series coupés and intended to give driver and passenger better support when the car was being driven the way its performance and suspension seemed to demand. All this was to the good; the bodykit was not universally liked, however, as it did look like a rather tacky add-on bought from an aftermarket supplier.

THE 535i FOR NORTH AMERICA, 1985

There are two reasons for the 535i's high pleasure quotient – its ability to go and its ability to stop. ... We love the 535i, but lament its high price.

Motor Trend, 1 March 1985

For the 1985 season, BMW replaced the 3210cc engine in its North American models by the 3430cc engine already seen in the 535i and M535i models for other markets. The new engine was of course simply a bigger-bore version of the one in the earlier US-market car and, like the version optional in Germany, it came complete with emissions-control equipment. For the USA, however, it was very slightly re-tuned.

One reason for replacing the 3210cc engine was, of course, to streamline production. However, it was important that the new engine should not do too much damage to the car's fuel consumption, and so for this reason BMW fitted the US-market cars with a much taller axle ratio of 2.63:1. The effect of the new engine was felt in better mid-range acceleration, but the US-model 535i was not at all the same car as the one which wore 535i badges elsewhere. To have offered it with the bodykit and sporting pretensions of the M535i would have been ridiculous, and so a Motorsport variant was never sold in the USA.

That the European and North American engines were not so very different in their power and torque outputs was an interesting illustration of the times. On the one hand, exhaust emissions regulations were about to be introduced in Germany and so BMW had to use power-sapping three-way catalytic converters in the exhausts of the European cars. On the other hand, engine management (this time by courtesy of the Bosch ME-Motronic system) had now reached a stage where good control of emissions could be achieved without anything like the power losses sustained by earlier engines with emissions control equipment. The engine in the North American 535i delivered 182bhp at 5400rpm as compared to 185bhp at the same crankshaft speed in European trim; those 3bhp had been sacrificed in the interests of greater torque, which in the US car went up from 209 lb/ft at 4000rpm to 214 lb/ft at the same crankshaft speed.

BMW North America once again marketed the 535i as a sports saloon, and all the cars had five-speed overdrive manual gearboxes and low-profile 60-section tyres on stylish alloy wheels. ABS was part of the standard equipment, but the desirable (some would argue essential) extra of a limited-slip differential was only available at extra cost. And cost was certainly a drawback of the North American 535i. Not that the customers minded; they were prepared to pay a premium price for a premium product, and sales once again reflected the fact that BMW North America had understood their market perfectly.

RUNNING CHANGES, 1985–86

Four new models plus a new version of one of them for the USA was enough for BMW to cope with in one 12-month period. So there was relatively little that was new on the company's stand at the Frankfurt Show in autumn 1985.

What showgoers did see were simply evolutionary changes to existing models. The 520i gained an extra 4bhp and 7 lb/ft of torque from the arrival of L-Jetronic electronic fuel injection in place of its K-Jetronic mechanical type. The revised engine was rather more high-revving than the earlier one, and this change in its characteristics led BMW to switch to a lower axle ratio for models with manual transmission and to shuffle the internal ratios of their five-speed gearboxes. The 520i was also given low-profile tyres for the first time at the 1985 Frankfurt Show. All this brought the performance of the smallest six-cylinder model uncomfortably close to that of the 525i, and so that car was in turn given lower overall gearing in order to keep its acceleration up to scratch.

More changes followed in February 1986, when the 520i and 524td were fitted with disc brakes at the rear. This left the 518i as the only model in the E28 range still to have a disc-and-drum system; all the others now had discs all round.

THE 524d, 1986

There were more evolutionary changes at the 1986 Frankfurt Show, when ABS was standardized on the 520i and 528i, and new tyres were specified for the 524td (that now switched to 195/70 HR 14s) and the 525i and 528i (that now had 200/60 VR 390s on the metric-sized alloy wheels already seen on the 535i). To suit the recently introduced legislation in Germany that favoured emissions-controlled cars, the 520i was now made available optionally with a catalytic converter. This meant that 'cat' models were available at three levels of the market, with the 520i, 525e and 535i. However, the 518i, 525i and 528i could not be bought with clean exhausts – and BMW never did make them available with catalytic converters during the production lifetime of the E28s.

Meanwhile, a non-turbocharged version of the 2443cc diesel engine had been introduced in 1985 for the 3 Series, in a model called the 324d. Here, its 86bhp and 110 lb/ft of torque gave quite a reasonable performance – although the car was of course some 120kg lighter than the bigger 5 Series. Sales were strong enough to ensure the viability of the engine as a production unit, and so BMW decided to see if a non-turbocharged diesel version of the 5 Series would find any buyers and introduced a 524d at the 1986 Frankfurt Show. The company had nothing to lose: sales of the 324d justified continued production of the non-turbocharged engine, and the E28s would in any case only be on sale for another 15 months, as they would be replaced by the third-generation 5 Series in January 1988.

The 524d was aimed fair and square at the diesel taxi market in Germany, where Mercedes-Benz continued to be the dominant force. To keep its price down, BMW stripped it of everything that could be considered optional or unnecessary – athough it did retain power-assisted steering. Options were available of course, if only to tempt economy-minded private buyers, and it was possible to order a 524d with ABS, alloy wheels with low-profile tyres, and a number of other extras.

It must be said that the 524d had sufficient torque to give the sort of acceleration needed in cut-and-thrust town traffic. The car also had the refinement and handling expected of a BMW. What it did not have, however, was outright performance. The 0–62mph (0–100kph) acceleration claimed by the factory was 18.5 seconds, and although the maximum speed of 102.5mph (165kph) was no disgrace, the car took a very long time to get there. BMW did not decide to abandon the diesel taxi market altogether, but the fact that the third-generation 5 Series introduced in 1988 did not include a non-turbocharged diesel model tells its own story....

4 The E28 Super-Saloons

BMW took their time before introducing a high-performance variant of the E28. The original E12 M535i had demonstrated that the company's own Motorsport division was capable of delivering the goods, but the early 1980s saw BMW soft-pedalling a little on high-performance cars. The effect was most marked in the USA, where performance seemed for a time to take second best to achieving volume sales. In the meantime, the aftermarket tuning companies stepped into the vacuum left by the absence of a high-performance factory-built BMW with some exciting confections of their own.

THE ALPINA B7S, 1982

Today, the Alpina name is synonymous with bespoke high-performance BMWs, and the cars that are built by this Bavarian company are not only approved by BMW but are actually marketed as a brand in their own right. But Alpina started out small, as the spare-time activity of Burkard Bovensiepen, whose family had made its mark through the Alpina typewriter business.

However, young Burkard Bovensiepen was not particularly keen on following in the footsteps of his forebears. His hobby was tinkering with cars, and he cut his teeth on some Fiats for his own use. Then in 1963, he designed a twin-carburettor conversion to improve the performance of the latest BMW 1500 Neue Klasse saloon. The word about this conversion quickly spread, and Bovensiepen suddenly found that he was getting more orders than he could reasonably cope with in a corner of his family's typewriter factory.

So in 1965, he set up Burkard Bovensiepen GmbH to produce performance-tuning parts for BMW saloons. Trading under the name of Alpina, he quickly built a reputation for his company. Alpina tuning parts found their way overseas, and as the knowledge spread that BMW made exceptionally well-engineered sporting saloons, so did the knowledge that it was Alpina who could make them go even faster and handle even better.

Alpina moved out of the typewriter factory and established itself at Buchloe, in the Allgäu region of southern Bavaria about 60 miles south-west of BMW's Munich headquarters. For the rest of the 1960s, the company operated as a supplier of aftermarket parts, but from 1969 things began to change. The company set about creating a more respectable image for itself by establishing its own racing team and then by developing complete conversion packages that matched blueprinted engines to uprated brakes and the like, in order to produce a properly engineered result.

Some, but by no means all, of the Alpina conversions from this period were actually built at Buchloe. The most popular base-car was the 2002, and after that ceased production in the mid-1970s, Alpina turned its attention to the new E21 3 Series cars. Conversions of the 630CSi coupé were also available.

The E28 Super-Saloons

Above all, BMW's own super-saloon version of the E28 was discreet. The cross-spoke wheels were special, and there was an M5 logo on the grille and bootlid, but who would have thought that this ordinary-looking saloon was capable of 153mph?

Alpina-tuned BMWs found a ready market in the USA during the 1970s, although that country's ever-changing exhaust emissions regulations did cause problems. Alpina shipped sets of standard parts out to various agents in the USA, only to find that the engines would not meet emissions regulations. Some agents, determined to make a return on their investment, modified the parts they had in order to meet the regulations, and in due course developed their own conversions which no longer needed Alpina's input.

Meanwhile, new regulations in Europe – covering issues such as drive-by noise levels – were also forcing Alpina to think about keeping closer control of the conversion work bearing its name. The happy result

The E28 Super-Saloons

The Alpina side decals seen on this car were optional extras, but the alloy wheels were a standard fitting.

was that the company decided in the late 1970s to pull out of the aftermarket conversions market altogether; to enlarge its Buchloe factory in order to bring all the conversion work back in-house; and to sell only completely re-engineered cars. That has been its policy ever since, although there has been one honourable exception to Alpina's rule of building cars in-house: it granted a conversion franchise for Britain to Sytner, the BMW dealers in Nottingham. This enabled the Buchloe factory to concentrate on left-hand drive cars while the British franchise holders looked after right-hand drive cars.

At the end of the 1970s, turbocharging offered the most promising way of extracting additional performance from existing engines without affecting exhaust emissions. Alpina's B7 Turbo conversion of the E12 5 Series (see Chapter 2) had pioneered that route to high performance with a 300bhp turbocharged version of the 3-litre big-six engine, and the company chose to go the same route for its first conversion of an E28 model.

Alpina's policy has always been one of gradual improvement, however, and so a new engine was developed for the E28 application. This was a turbocharged version of the 3.5-litre big six which featured as a production engine in the 6 Series coupés and 7 Series saloons. This engine punched out 330bhp and 332lb/ft of torque and was known as the B7S conversion. It was available in both the E28 5 Series and the E24 6 Series coupés. Brakes and suspension were of course uprated to match this quite awesome power that promised 0–62mph (0–100kph) in just 5.8 seconds and a maximum speed of 162mph (261kph).

There was no doubt that the B7S was capable of giant-killing performance, but it was not a very refined car. Like so many turbocharged engines of its time, the Alpina engine made a fierce transition from off-boost to boosted performance, and the sudden kick in the back that accompanied that transition could catch the driver unawares. The B7S needed a good driver to tame it, and that inevitably limited its appeal.

THE ALPINA B9 3.5, 1983

> A gorgeous supercar with phenomenal performance and handling to match.
> *Drive*, March 1985

Alpina started putting BMW's big 3.5-litre six-cylinder engine into the E28 5 Series saloons more than two years before BMW did it themselves with the 535i and M535i. Moreover, the Alpina conversion, which was known as the B9 3.5, had an Alpina-tuned engine which gave awesome performance that outclassed BMW's own, much milder, efforts. With 0-60mph acceleration of 6.4 seconds and a top speed of 153mph, the B9 was in the same league as high-performance sports exotics like the Ferrari 308GTB or Porsche 911SC – and it remained a practical five-seater saloon with a proper boot.

The basis of the B9 was an E28 528i with the five-speed manual gearbox and a limited-slip differential. Into that went the 3453cc engine from the 6 Series coupés and 7 Series saloons, but modified from its standard 218bhp to give 245bhp at 5700rpm with a modest increase in torque to 231lb/ft at 4500rpm. The Alpina modifications consisted of a new cylinder head with hemispherical combustion chambers and larger inlet valves, new pistons to give a higher compression ratio, a high-lift camshaft and a dual exhaust system.

The standard braking system was adequate to cope with this power, but the car's suspension was considerably modified to give a more sporting feel and to cope

	M5 Performance				
Model	0-60mph 0-100Kph	50-70mph 80-110Kph	Max speed	Average mpg	Source
UK Spec	6.3sec	6.8sec	153mph 246kph	18–19mpg/16–15/100km	BMW
US Spec	6.7sec	-	150mph 242kph	10mpg/28/100km (EPA City)	BMW

with the way drivers were likely to use its new-found performance. So the B9 had stiffer progressive coil springs and Bilstein dampers. These were combined with wide Pirelli P7 tyres on Alpina's characteristic spoked alloy wheels; there were 205/55 VR 16 tyres on 7-inch wheels at the front, and 225/50 VR 16s on 8-inch rims at the rear.

The B9 also carried Alpina's own deep air dam at the front and the company's boot lid spoiler. Alpina decals were applied to the flanks and to the front air dam, although such 1970s addenda were increasingly becoming associated with 'boy-racer' cars and were really out of place on a car aimed at an older, wealthier clientele. The bright deckchair striping of the velour seats also seemed unfashionably dated, but the leather-bound steering wheel with its smaller-than-standard diameter and the numbered builder's plaque on the facia were reassurances that this car did have a thoroughbred pedigree.

Buyers had to pay a lot for the privilege of owning a B9, which ensured that the cars retained their exclusivity. They were built in small numbers, too. In Britain, where the cars were assembled by Alpina's franchise holders, Sytner, no more than 40 were expected to find customers in the first year of their availability. That was not surprising: a B9 cost nearly £23,000 at a time when a 528i without extras cost £13,575!

THE ALPINA B2.8, 1983

> One of those cars which could be taken by the scruff of the neck and flicked through the most intricate of twisty bend sequences in complete confidence.
> *Performance Car*, November 1983

Not long after introducing the B9 3.5 into Britain, it became apparent to the Sytner franchise that there was a market for a less expensive conversion. Rover's 190bhp Vitesse had been selling strongly since its 1982 introduction, and it was clear that the market was ready for high-performance cars in this price bracket. At £14,950, the Vitesse was something of a bargain, and there was no way that an E28 could be given the full Alpina treatment and remain competitive on price. However, Sytner believed that the top-model 528i could be given a more sporting demeanour with the

---

M5 (1985–87)

As for 535i and M535i (see Chapter 3), except:
Engine: 3453cc six-cylinder (93.4mm bore x 84mm stroke) with four valves per cylinder, two overhead camshafts, 10.5:1 compression ratio and Bosch Motronic electronic fuel injection; 286bhp at 6500rpm and 245lb/ft at 4500rpm.
  US models equipped with three-way catalytic converter; 256bhp at 6500rpm and 243lb/ft at 4500rpm.
Transmission: Five-speed overdrive gearbox (ratios 3.51:1, 2.08:1, 1.35:1, 1.00:1, 0.81:1); no automatic option. Axle ratio 3.73:1 or (US models) 3.91:1.
Tyres: 220/55 VR 390 TRX on 165 TR 390 alloy wheels, or 225/50 VR 16 on 7.5J x 16 alloy wheels.
Unladen weight: 1470kg (3240 lb). US models' kerb weight 3420 lb (1551kg).

The E28 Super-Saloons

Multi-spoke alloy wheels and distinctive spoilers mark this E28 out as an Alpina variant.

addition of some items from the Alpina range, and that buyers would be prepared to pay a little over the odds for the prestige of the Alpina name. So Sytner developed their own high-performance version of the E28 with the full approval of the Buchloe factory. They called it the Alpina B2.8.

Each B2.8 conversion started life as a right-hand drive 528i with manual sunroof and limited-slip differential. Sytner then exchanged the standard exhaust for the B9's twin system, which released a little more power to give 192bhp instead of the standard 184bhp output. An Alpina suspension kit, identical to that on the B9, was the next step; and the wheels and tyres were also exchanged for B9 items. The B9's deep front air dam, boot lid spoiler and decals were added, and there were of course unique B2.8 badges on the grille and boot lid. Inside, though, the car remained standard except for an Alpina gear knob and a leather-rimmed Alpina steering wheel.

The B2.8 may not have accelerated any faster than a standard 528i, and its maximum speed of 132mph was pretty much the same as that of the standard car. However, it was certainly a more entertaining car to drive, with much more sporty handling than the standard car and a superb high-speed stability. Like all the

The E28 Super-Saloons

Colour made a big difference to the M5's appearance. This white car is a South African-built version, and the M-Technik bodykit of sills and spoilers is much more apparent than on the German-built car pictured on p.75.

other Alpina BMWs, it was of course exclusive. When new, it cost around £16,595 in Britain, or just over £1,600 more than the already expensive Rover Vitesse. Sytner also expected to make no more than three dozen examples each year – and probably made fewer than that.

THE HARTGE H5S

> Just think of it as a four-door Porsche 911.
> *Car and Driver*, March 1984

Herbert Hartge GmbH is based at Beckingen in Germany, and for many years has specialized in performance enhancements and cosmetic accessories for the BMW range of cars. For the E28 range, it offered both engine upgrades for the 528i and a complete (H5-3.4) engine change to the big 3430cc six-cylinder engine first seen in the 635CSi and 735i during 1982. Its range included upgraded exhausts, suspension improvement kits, bodykits, and alloy wheels. In addition, items such as alloy footrests and alloy pedals were (and still are) available for the E28 5 Series.

The H5S conversion was based on the 528i, and featured the 3430cc engine that was essentially the same as the one that BMW itself would later use in the 535i models. As manufactured by BMW in the early 1980s, it was rated at 218bhp at 5200rpm and 224lb/ft of torque at 4000rpm. However, the Hartge version had a mildly modified cylinder head, a different camshaft, tubular exhaust headers and a low-restriction exhaust system to give 237bhp at 5400rpm. US versions were offered through Performance Plus, Inc. in Dania, Florida, and had a catalytic converter in the exhaust that was estimated to rob the engine of about 10bhp.

The engine transplant was only part of the H5S package, however. Hartge also offered lowered and stiffened suspension with Pirelli P7 tyres (P205/55VR16 at the front and P225/50VR16 at the rear) on alloy wheels. The chassis could be tautened by a reinforcing member bolted between the front suspension towers, and front and rear spoilers could be added to improve high-speed stability as well as the car's looks. Side stripes were available to make the car distinctive, and Recaro front bucket seats were on offer, together with a high-powered sound system.

The H5S was a more overtly tuned car than its rivals from Alpina, with an exhaust note that betrayed its purpose and a ride that was much firmer than the standard BMW offering. Maximum speed was approximately 142mph (230kph) in European trim, and 60mph came up from rest in just 6.3 seconds.

THE M5, 1984

> Very pricey indeed in terms of perceived value for money, but as the ultimate upper-crust 'Q car', the BMW M5 probably has no peers.
>
> *Motor Sport*, October 1986

There had been Motorsport-developed 5 Series saloons since the arrival of the E12 M535i back in 1979, and of course in 1984 the tradition had been continued with an E28 model carrying the same designation. However, no-one in BMW or in the Motorsport division itself can have been in any doubt that the basic 5 Series car was capable of much more performance development, or that such a high-performance car would find a ready market.

So the Motorsport division set to work to develop the ultimate E28 derivative that would go into production with the M5 designation lending it the exclusivity already associated with the M1 supercar available between 1978 and 1981. That exclusivity would be enhanced by the limited availability of the M5 that would be hand-built to order by the Motorsport division. In the beginning, a cautious 250 examples a year were planned for worldwide consumption, although demand soon led BMW to raise that figure. Between the start of M5 production in October 1984 and its end in December 1987, a total of 2,145 M5 cars were assembled at Garching, and a further 96 cars were shipped abroad for overseas assembly. (The last car was not actually built until June 1988.) That total of 2,241 cars averages out at more than 700 a year over the production run, although in fact production was slower than that in the beginning and the numbers were boosted by no fewer than 1,235 cars for the US market at the end of 1987.

For this bespoke E28, there was an obvious choice of engine right from the beginning. The Motorsport division had been steadily developing the big-block six-cylinder, which by the early 1980s was sporting twin overhead camshafts, four valves per cylinder, Bosch Motronic engine management, and the M88 designation. A reshuffle of bore and stroke dimensions had also left it with a capacity of 3453cc rather than the 3430cc of the 3.5-litre two-valve engine. With 286bhp at 6500rpm and 250lb/ft at 4500rpm, it was far and away the most powerful engine in the BMW stable, and there was no doubt that it could not simply be dropped into a standard E28 and left to do its worst. Some complementary development of brakes and suspension would be needed to ensure that this power could be used safely and responsibly.

The first M5s were built in October 1984, and filtered out to VIP customers before the official announcement of the new model. This was originally planned for the Geneva Show in March 1985, but when it became clear that the catalyst-equipped version would not be ready in time and that Swiss ecological sensibilities would be offended if a non-cat car were to be shown, BMW changed its collective mind. So the public unveiling was at the Amsterdam Motor Show a month earlier, in February 1985.

The show car was used to gauge public reaction, and BMW paid more than its usual attention to customer feedback. As the M5 was to be a determinedly expensive car, selling to a wealthy and notoriously fickle clientele, BMW was not going to sell many if it got the specification wrong. The company already knew that M5 buyers would not appreciate an extrovert appearance such as the M535i had; on the contrary, they wanted discreet Q-cars. To that end, the show car had steel wheels and no spoilers, but feedback revealed that to be rather too much understatement for the market. So when the production cars came along in 1985, they had discreet spoilers and distinctive alloy wheels.

These features helped the M5 to be distinguished from other members of the E28 clan – though it took a knowledgeable eye and a close look. For a start, the car had a deeper front air dam that incorporated long-range driving lamps beside the turn indicators, and the ribbing at the bottom of this was carried through the shallow side skirts to the lower trailing edge of the rear wings. There was a boot lid spoiler, too, but it was painted in the body colour like the door mirrors, and was barely noticeable at first glance. Just two features gave away what the car really was. The more noticeable of these was the BBS cross-spoked alloy wheels (that to the uninitiated simply looked like a rather attractive item that might have come from the 5 Series options list), and the less obvious was the discreet M5 badges on the grille and boot lid, each one bearing the angled Motorsport tricolour flash.

The interior was equally discreet, but the M5 logo on the kick-plates made clear on boarding the car that this was no ordinary 5 Series BMW. There were velour

Performance Comparison

N.B. These figures are for European-specification cars and are intended as a guide only.

Model	Power bhp	Torque lb/ft	0-62mph	Max. speed mph
Alpina B7S	330bhp	332 lb/ft	5.8 (*)sec	162mph/261kph
Alpina B9 3.5	245bhp	231 lb/ft	6.4 (*)sec	153mph/246kph
Alpina B2.8	192bhp	173 lb/ft	7.6(*)sec	132mph/212kph
Hartge H5S	237bhp	N/A	6.3 (*)sec	142mph/228kph
Alpina B10	261bhp	240 lb/ft	6.1 (*)sec	150mph +/241kph+
BMW M5	286bhp	245 lb/ft	6.3 (*)sec	153mph/246kph

carpets, matched by velour kick-panels on the doors and velour on the parcels shelf, while the door inserts were in Highland fabric. The seats were special, too, with close-fitting power-adjusted bucket seats at the front and heavily-contoured seats for two in the rear. Leather was standard for the upholstery and certain trim elements, but in its place buyers could specify natural buffalo hide. Recaro front bucket seats were also an option. In all cases, though, the seats carried a small metal Motorsport tricolour flash on their backs.

Ahead of the driver, a tricolour band was applied to the central spoke of the three-spoke leather-bound steering wheel, and there was a discreet Motorsport logo on the rev counter. The 6900rpm redline on this instrument, and the 260kph (or 160mph) speedometer were the only other indications that this was a very special 5 Series indeed.

The Motorsport engineers had allowed themselves some self-advertisement under the bonnet, however, as well they might. The 24-valve engine was beautifully presented in a tightly-packed engine bay, with crackle-black finish setting off the cast alloy cam cover with its 'M Power' legend. The complicated valvetrain gave this engine a rather harsher, more mechanical note than the other BMW sixes of the time, but this was entirely in keeping with the sporting nature of the car. Astonishingly for such a highly tuned engine, the M88 was also extremely flexible, with enough low-speed torque to give entirely untemperamental behaviour in city traffic. This was largely responsible for endowing the car with its delightful dual nature; on the one hand, the M5 could be used as an everyday family saloon, and on the other, a determined thrust on the accelerator pedal could transform it into a road rocket with turbine-like acceleration.

The car's standing-start time to 60mph of just over 6 seconds (and some road-testers claimed even better times) was in the supercar class for 1985, and the maximum of 150mph (240kph) (again bettered in some tests) all that anyone could reasonably want of a roadgoing four-door saloon.

All those 286bhp drove through a reinforced Getrag 280 five-speed gearbox with a conventional gate arrangement, and were fed to the rear wheels through a 25 per cent limited-slip differential. The suspension was lowered and equipped with beefier anti-roll bars, while its stiffened coil springs backed up by monotube gas dampers gave a ride that was noticeably firmer than standard. There was limpet-like grip from the fat Michelin TRX tyres with their metric wheel dimensions (though in some markets more conventional Imperial-sized wheels and tyres were used) and, although it was not impossible to provoke some oversteer in the wet, the car's handling was beautifully balanced in normal driving. This balance was aided by careful weight distribution: to improve it, the battery had been relocated in the boot. Under fierce acceleration from rest, it was possible to provoke both wheelspin and a certain amount of vibration from the rear subframe which some road tests described as axle tramp, but at high speeds the car was reassuringly stable. The brakes, with ABS as standard and ventilated discs with extra-large callipers at the front, gave amazingly consistent stopping power that positively encouraged the full use of the car's performance.

It was some time before right-hand drive versions of the M5 became available, and in Britain the car made its bow during July 1986. Just 177 examples were made for this market in the 17 months of its

The E28 Super-Saloons

Under the bonnet of the M5 lurked the 24-valve Motorsport engine.

availability, although they were not the only right-hand drive M5s: others were assembled in South Africa from CKD kits in 1987 and 1988. The UK market cars all had the electric sunroof that was only optional on the German-market models, and they came with Pirelli P700 tyres in 225/50 VR 16 size, once again on BBS alloy wheels. ABS, electrically adjusted front seats, air conditioning, central locking, electric windows, an on-board computer, heated washer jets, a heated driver's door lock and a rear window blind were all included in the impressive standard specification. A repositioned brake master cylinder was also a feature of the right-hand drive cars, as BMW found that simply switching it from the left side to a corresponding position on the right gave poor braking feel.

The USA had to wait even longer for this ultimate E28 that finally crossed the Atlantic in August 1987. BMW anticipated a big demand, and references to limited availability were accompanied by suggestions that no more than 600 would be available to US customers during 1987 – more than twice the original projected annual figure in just five months! BMW felt that these volumes would not compromise the car's exclusivity in such a big country, and were only too pleased to oblige by building more M5s when demand went through the roof. However, they had reckoned without the litigation-happy American public. Owners of the first M5s banded together and sued the company for false representation when they discovered that their cars were not after all as exclusive as they had been led to believe when they bought them. BMW got out of that one nicely by giving each of them a credit note towards a new BMW ...

The M5 carried the standard US-specification impact bumpers of the time, and it was available only in black with dechromed bumpers, windscreen and window surrounds, and tail lamp trims; to that extent, it was not a bespoke model like the M5s for other markets. A sunroof, air conditioning, cruise control and leather upholstery were all standard, and, of course, the M5's exhaust emissions had been cleaned up to comply with US regulations. So the car offered to American buyers had a three-way catalytic converter in its exhaust which, together with other emissions-related work, robbed the engine of a massive 30bhp. Torque, at 246lb/ft, was much less affected.

Even so, some very careful work had been needed to make the Federalized M5 into the same sort of super-saloon as the car sold in other markets. The Motorsport division had lowered its final drive to get performance close to the levels of the European M5. The results were impressive, because the US version lost only half a second in the 0-60mph sprint and still had a claimed maximum of 150mph. The downside was its fuel consumption, which was most definitely not for the faint-hearted. In fact, it was so bad that the US M5 was subject to the 'gas guzzler' tax imposed on new cars with spectacularly poor fuel economy, and as a result buyers had to pay $2,250 on top of the showroom price. BMW knew that this would not be a deterrent, though: the sort of buyer who could afford an M5 in the first place was unlikely to be troubled by such a trifling amount!

THE ALPINA B10, 1987

> Sounds like a military jet and, come to think of it, performs like one, too.
> *Performance Car*, April 1987

It took the tuning specialists some time to come up with a car that could offer more than the M5, but Alpina knew that it had to if it were to remain in business. The B7S conversion certainly did offer more power and more performance than the M5, but Bovensiepen was not happy with the poor drivetrain refinement of the turbocharged engine. The M5 was an altogether more sophisticated piece of machinery, and its dual nature gave it a much wider customer appeal than the B7S.

Alpina's solution was to develop the 3.5-litre engine just a little further and to match it to an automatic transmission – an option not available on any other high-performance E28. The car was called the B10 and was announced in 1987, just a year before the new E34 models were announced. That inevitably limited the number of E28 versions of the B10, but of course Alpina had not developed their new engine only for the 5 Series; there were also B10 editions of the 6 Series coupés and the recently superseded E23 7 Series cars, so that the engine and the B10 designation lived on beyond 1988.

With the B10 conversion, Alpina were

also able to offer keener pricing; basing the car on an M535i that already had the 3.5-litre engine meant that it was no longer necessary to buy the basic 3453cc power unit separately. The Alpina conversion boosted power to 261bhp at 6000rpm, while torque went up to 240lb/ft at 4000rpm. This was a lower-revving engine than in previous Alpina 5 Series cars, and it was also lower-revving than BMW's own M5. Clearly, Alpina had been aiming at refinement this time around, and the use of the switchable ZF four-speed automatic transmission confirmed the B10's mission.

As the B10 was based on the M535i, it retained that car's spoilers. However, the usual Alpina suspension improvements were incorporated, along with the company's trademark spoked alloy wheels shod with Pirelli low-profile tyres. It carried B10 badges on the grille and boot lid for easy recognition, but other Alpina touches were confined to the leather-bound steering wheel, the gear knob, and the builder's plaque on the dashboard.

Even though the B10 engine did not offer as much outright power as the Motorsport four-valve in the M5, it did bring M5 performance to an E28 5 Series with automatic transmission. Alpina claimed 6.1 seconds for the 0–60mph standing-start (as compared to BMW's claim of 6.3 seconds for the M5), and a top speed of 'over 150mph' (as compared to 150mph for the M5). The ride was as firm as Alpina customers had come to expect, and the road roar from the big tyres seemed more intrusive than in the M5, but there was no doubt that the B10 represented a real achievement for Alpina. The automatic transmission offered effortless through-the-gears performance and made city traffic less of a chore than in earlier Alpina models. In this, it was obviously intended to emulate the superb flexibility and dual nature of the M5. There was just one drawback to the B10's use of automatic transmission: the ZF box tended to change down on slow corners, so the driver had to be careful not to accelerate again too hard if he was not to make the tail step out of line.

Epilogue

Over the seven years of the E28's production run, the performance of the fastest roadgoing models reached levels that had simply been unimaginable in the heyday of the old E12 M535i. The 60mph standing-start could now be achieved in 6 seconds, and maximum speeds of 150mph or more were common. The B7S, despite its inherent lack of refinement, could reach 60mph from rest in under 6 seconds and storm on to a maximum of 162mph, but this was not the way the high-performance 5 Series would go when the E34 models came on-stream. BMW's own M5 had offered an unbeatable combination of performance, driveability and refinement, and even though it was not the fastest of the high-performance E28s, it did set the standard for the future.

Bertone's lovely 3200CS coupé helped establish the BMW style. Note the angled trailing edge of the rear window, seen on every 5 Series model.

(Right) The first-generation E12 cars were distinguished by flat bonnets.

(Left) The 02 range, developed from the Neue Klasse four-door saloons, played a major role in establishing the BMW sporting image.

(Left) *American sales were important to BMW, and the company went to great lengths to prepare its cars for that market. This is an E12 model.*

(Right) *The 5 Series cars were also assembled in South Africa. This is an E28 M5 model.*

(Left) *By the time of the E28 second-generation cars, the big bumpers needed on US models had been better integrated into the overall design.*

(Above) *Sporting handling and a distinctive appearance were the essence of the E28 M535i, which was mechanically the same as the ordinary 535i model.*

(Below) *Something about the stance of this US-market E28 makes its sporting character clear, despite the heavy bumpers.*

(Above) *The powerplant under the bonnet of both the ordinary 535i and the M535i was a typically neat and well-packaged affair.*

The style of the E28's dashboard was very much of its time. Note that this car has the selectable-mode automatic transmission.

The first M5 was introduced in 1984 and was based on the E28 second-generation 5 Series. Its discreet appearance could not disguise the purposeful air of this high-performance saloon.

This US-market publicity picture shows the E34 range, complete with the Touring model introduced in 1991.

The E34 was arguably the best-looking of all the 5 Series cars, and is seen here in US-market saloon form.

Alpina has created some hugely desirable high-performance versions of the 5 Series. This is an E39 B10 V8.

BMW's own high-performance E34 was the M5, initially powered by this 315bhp, 3,535cc straight-six engine.

Alpina used twin turbochargers to get 181mph and 0–62mph in 5.2 seconds out of an E34 5 Series. The car was called the B10 Bi-Turbo.

Cross-spoke alloy wheels add to the appeal of this US-market E34 saloon from the early 1990s.

BMW unveiled the M5 version of the E39 two years after the other models, but it was not available for several months after that.

The fourth-generation E39 car had beautifully curvaceous styling, but somehow looked much heavier than its E34 predecessor.

(Above) An in-dash navigation system was available for the E39 models.

5 Third Generation: The E34 Models, 1988–96

The E34 models were announced in January 1988, their first public appearance being at the Detroit Motor Show that month. The choice of a US venue for the worldwide launch was a clear demonstration of the importance to BMW of North American sales of the new range.

Not surprisingly, the new models had some very close similarities to the acclaimed E32 7 Series range that had been introduced two years earlier. In Britain, *Motor* magazine headed its report on the new cars 'Son of Seven', and there was no doubt that the cars looked a lot like their bigger sisters. The rather angular look that had characterised the E12s and E24s had been swept away in favour of a more rounded, aerodyamic shape. It was to some extent a scaled-down version of the hugely successful 7 Series design, but it also worked very well in its own right. The E34 undeniably had a presence which the earlier 5 Series cars had lacked, with restrained rather than strictly conservative styling. It was certainly the best-looking 5 Series range yet – and it was arguably not bettered by its own successor seven years later.

The E34s were slightly larger than the E28s which they replaced, being 100mm (3.9in) longer and 51mm (2in) wider. However, there had been a more fundamental change in proportions, too: the wheelbase was a full 136mm (5.35in) longer than before, allowing extra length in the passenger compartment and putting the wheels nearer the corners of the car to improve ride comfort. The E34s were also a shade (3mm, 0.1in) lower than the E28s.

The increased width, that allowed both for more passenger space and for wider tracks to give greater stability, of course worked against the aerodynamic improvements made in other areas by creating a larger frontal area for the car. Nonetheless, BMW claimed a Cd of between 0.30 and 0.32 for the first models, which made them up to 50 per cent more aerodynamic than the older cars. The more tapered profile with its low nose and high tail was fundamental to this improvement, as was the steeper rake of the windscreen and rear window. However, detail changes also made their contribution. On the one hand the windscreen and side windows were less inset than before (and here it must be remembered that Audi had shown the way with flush glazing on their medium-sized saloons). On the other, there were spoilers subtly integrated into the body shape. The small front air dam was barely noticeable, and the shaped trailing edge to the boot lid looked more like the work of a stylist than that of an aerodynamicist.

Computer-aided design had enabled BMW to make the new bodyshell considerably stiffer than its predecessor with an increase of just 5 per cent in weight before running components and

Third Generation: The E34 Models, 1988–96

Still recognisably BMW, the driving compartment of the E34 was a step forward from earlier 5 Series types.

trim were added. There was a dynamic improvement of 30 per cent in both bending and torsion, while the static figures were even more impressive and showed improvements of 43 per cent in bending and 70 per cent in torsion. This had been achieved by using stiffened joints at all major junctions (such as between the floor and the body pillars), by using stronger sills, and also by bonding the windscreen and rear window to the body. The central passenger cell was now strong enough to remain largely intact in a 35mph barrier crash, which was 5mph above the usual requirement. And BMW had no intention of allowing corrosion to lower the bodyshell's resistance to impact, either, because approximately 45 per cent of it (measured by weight) was zinc-coated.

Earlier 5 Series cars for the USA had needed ugly 'impact' bumpers to meet regulations which demanded that a 5mph impact should not damage safety-related items on the car. By the time the E34s were announced, these regulations had been relaxed and stipulated resistance to a 2.5mph collision. So BMW were able to integrate the US-mandated collision-resistant bumpers into the basic design and, in order to streamline production and save costs, they standardized these for all markets. Thus the demands of the US market did not ruin the svelte new styling, and the undoubted benefits of these bumpers were brought to other markets where they gave the new BMWs an extra competitive edge.

The new bumpers had plastic coverings and a black rubber facing that neatly matched the bump-strips running along the cars' flanks. However, underneath they consisted of lightweight aluminium alloy bars capable of absorbing impacts of 2.5mph without damage to safety-related items. These were mounted on collapsible box-section tubes that would absorb impacts of up to 9mph before transmitting the collision forces into the main body structure. These tubes were easy to replace, which of course lowered repair costs after a collision.

Inside the car, the improvement in passenger space was quite marked. A more rounded facia contributed to a more luxurious feel and betrayed the now familiar BMW attention to detail. As on the E32 7 Series, its upper half was made of soft-feel plastic that reduced by 35 per cent the fogging effect on the windscreen caused by plasticizer emissions. There was telescopic adjustment for the steering

Third Generation: The E34 Models, 1988–96

column on all cars, with a four-spoke wheel on all models except the range-topping 535i that had a three-spoke leather-rimmed type. Standard right across the range was a new heater with separate rotary controls for the right and left sides of the car and automatic temperature regulation that took account of both the interior temperature and that outside the car. Automatic air conditioning was an optional extra. Electronics were again behind the warning-message system (known as Electronic Check Control) fitted to the instrument panel of the mid-range 525i and all the more expensive models. This monitored a wide range of the car's systems and, when appropriate, displayed a brief message to warn of any one of 23 problems or malfunctions – and it could do so in any one of six languages to suit the market the car was intended for.

Again as in the 7 Series, the rear seats were fitted as standard with a third three-point inertia-reel belt for a centre passenger. Large quantities of sound-proofing insulated the cars' occupants from mechanical and road noise, while wind noise was banished by new three-section door seals and by new assembly methods (using laser measurement and surface scanners) that allowed very precise fitting of the doors. The overall result was that the E34 models were remarkably quiet cars. aided by engine insulation on petrol

The E34 was a good-looking machine from any angle. This picture shows an early example - actually a South African-assembled 535i - with the narrow grille used between 1988 and 1994.

Third Generation: The E34 Models, 1988–96

On the third-generation cars, the impact-absorbing bumpers and marker lights required in the USA were very successfully integrated into the basic design.

models and encapsulation on diesels.

One very useful option was a split back to the rear seat, although this was not made available before the autumn of 1988. Either section, divided one-third and two-thirds, could be folded down to create an extra three feet of space for long loads to be fed through from the boot. And the increased stiffness of the bodyshell had allowed BMW to make another very worthwhile improvement by providing the boot with an unusually low loading sill.

All the E34s had central locking with deadlocks, and as the years went by this anti-theft feature would be reinforced by many others. Special care had been taken in designing the electrical circuits (not least because the cars depended so heavily on electrical and electronic systems) that had high-quality positive connectors to ensure reliability. Despite their conventional round appearance, the headlamps were of a new ellipsoid design that gave superb lighting with excellent beam control. All models had green-tinted heat-filtering glass with the radio aerial incorporated in the rear window demister element. Attention had been focused on the windscreen wiper arrangements, too. The smaller-engined cars had a small

---

The BMW Family Touring, 1994

In June 1994, BMW made some publicity capital out of a prototype Family Touring car. Specially designed for owners with children, it had a number of special features. These included integrated child seats which were adjustable to suit the size of their occupants, fold-out tables, reading lights, a fixed central stowage box and a second removable box for toys and other paraphernalia. These boxes were also plumbed into the car's electrical system, and the fixed box could be used as a food or drink heater while the removable one doubled as a cooler.

Third Generation: The E34 Models, 1988–96

Wheels could make a considerable difference to the appearance of the E34s. This example, a South African-built 535i, has smart alloy wheels with radial spokes ....

... while the sporty-looking cross-spoke alloys here are fitted to an entry-level 518i for the UK market.

aerodynamic wing on the driver's side wiper to prevent lift-off at speed, while the 530i and bigger-engined models had the latest 7 Series technology of an automatic increase in the blade pressure.

THE 520i, 524Td, 525i, 530i AND 535i, 1988

The five models of the E34 5 Series announced in January 1988 all had six-cylinder engines. All these were developments of the engines in the superseded E28 models, although the mid-range 525i now had a derivative of the M20 small-block six rather than the M30 big six seen in the earlier 525i. Petrol-engined models for all markets came as standard with a three-way catalytic converter and Lambda sensor.

The lowest-powered of the five was the turbodiesel 524td, with 115bhp but more torque than before thanks to its new Bosch

101

Third Generation: The E34 Models, 1988–96

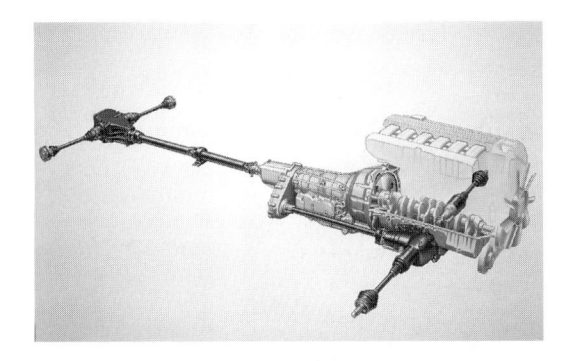

The 4x4 drivetrain of the 525iX models had been seen earlier in the smaller 3 Series cars.

DDE (Digital Diesel Electronics) control system. The 2-litre petrol engine in the 520i came with 129bhp, and the stroked and bored 2.5-litre in the 525i boasted 170bhp. There was no longer a 528i; instead, BMW fielded a 530i with a 188bhp 3-litre engine, and at the top of the range was a 211bhp 535i with the 3.4-litre engine from the 7 Series. There was no M5 yet; that was scheduled for later introduction, as was a four-cylinder 518i.

Transmissions were carried over from the previous models, with five-speed overdrive manual gearboxes standard across the range and four-speed overdrive automatics the only option. There was power-assisted steering across the range, but instead of the recirculating-ball type fitted to the smaller-engined cars, the more expensive models had a ZF Servotronic system. This could be fitted optionally to the lesser models. It was a speed-sensitive system that gave high assistance at low speeds and progressively less as the car's speed increased.

The basic suspension layout was as before, too, with MacPherson struts and double wishbones at the front, and semi-trailing arms at the rear. However, detail attention had banished the oversteer of earlier 5 Series cars. Optional on the rear was a hydropneumatic self-levelling system to compensate for full loads in the boot, and the 7 Series had inspired a second option of electronically controlled dampers with comfort and sports settings. The dampers were made by Fichtel and

---

The Natural Gas E34

Since the late 1970s, BMW had been working on a zero-emissions vehicle, not least because legislation threatened for the end of the century in California demanded that a certain percentage of all road vehicles would fall into this category. One of the strands of the programme was the development of a car that would run on pure hydrogen (that can be produced from water by electrolysis through solar or electrical power), and a stage in the development of this was engines that would run on natural gas as well as on petrol.

In December 1995, BMW announced the availability in Germany of a natural gas-powered variant of the E34 range, called the 518g Touring. It was released alongside a 316g Compact version of the E36 range. Its engine was essentially the existing 518i type, but with changeover valves to allow it to run on compressed natural gas stored in an extra fuel tank. BMW claimed that the changeover from petrol to gas and vice versa could be made seamlessly on the move, and that the car's fuel consumption on gas was the same as on petrol. The advantages were 20 per cent less carbon dioxide and 80 per cent fewer hydrocarbons in the exhaust gases.

The 518g Touring was never made available outside Germany because no other country had the necessary refuelling infrastructure. It was also substantially more expensive than the standard vehicle, and was not available for very long.

Third Generation: The E34 Models, 1988–96

This is a UK-market 525i Sport, which came equipped with a bodykit of side sills, rear apron, rear hoop spoiler and an additional lip on the front air dam.

Sachs, and their settings were changed by electrically powered rotary valves within the dampers themselves. Standard bump and rebound control on all models was by twin-tube gas dampers.

All models had a 10-inch vacuum-operated brake servo that superseded the hydraulic servo used on some of the E28 models, and ABS was standard on every model except the 520i, where it was an extra-cost option. On the 535i, ASC wheelspin control could be fitted at extra cost. The E34s also had bigger brakes than before, with 302mm (11.89in) discs at the front and 300mm (11.81in) discs at the rear. The slower models – 520i and turbodiesel 524td – had solid front discs with a width of 12mm (0.47in), but the additional performance and weight of the other models was matched to ventilated front discs with a width of 22mm (0.87in). Room for these larger-diameter brakes was made by larger wheel rims with a 15-inch diameter, while the rolling radius of the wheels was kept much as before by the standardisation of lower-profile 65-section tyres. Wider metric-sized alloy wheels were optional, with even lower-profile 45-section tyres.

The 540i Sport

The US market was treated to a special run-out edition of the E34 range in 1995. This was called the 540i Sport, and was effectively a hybrid of the 540i and M5. The car had the standard 540i mechanical specification, but was turned into an M5 lookalike with the Motorsport car's body addenda and wheels. It also boasted the M5's special suspension and brakes. Just 200 were built.

Third Generation: The E34 Models, 1988–96

The 525i quickly became the best-seller of this initial range, although it was actually outsold by the 520i at first. In descending order, sales success then went to the 535i, 524td and 530i – and in fact sales of the 530i dropped off so sharply that the model was dropped from the range after just two seasons. Most probably, it was the car's now-elderly big-six engine that discouraged the buyers. It endowed the 530i with poorer acceleration than a 525i, and it had a lower top speed than the smaller-engined car. The 525i was slightly thirstier, but this drawback was offset by its diet of Normal Unleaded petrol instead of the more expensive Super Unleaded demanded by the 530i.

While all the 530i and 524td models were built in Germany, BMW's South African plant also began making 520i, 525i and 535i models during 1988. Only 570 cars were built in the 1988 calendar-year, of which the majority were entry-level 520i types. However, as production increased to level out at around ten times that number annually, the 525i took over as the best-seller and the 520i dropped to second place.

THE 518i, 1989

BMW's plan in marketing the E34 models had been to score against Mercedes-Benz by offering six-cylinder engines right across the range. The Stuttgart company's medium-saloon offerings had four-cylinder engines in the 2-litre (122bhp) 200E and 2.3-litre (136bhp) 230E models, and the smallest six-cylinder was the 166bhp 2.6-litre in the 260E. In the two manufacturers' native Germany, it was important to keep up appearances – but in export territories it was more important to promote sales. So BMW prepared a four-

The Touring models were introduced in 1991, and brought svelte styling to the estate-car market. The twin sunroofs are clearly visible in this picture of a car at speed.

cylinder 518i for certain export markets and put it into production during 1989. Among the markets where the car was sold was Britain, where it was introduced in May 1990.

The 1766cc engine that had been used in the last of the E28 518i models had gone out of production altogether in 1987, to be replaced by a new and more powerful 1796cc engine. This had first made its appearance during 1987 in the E30 318i, and in the E34 518i it had the same Bosch Motronic engine management system and the same 113bhp tune. The 518i was no ball of fire – though it did accelerate faster than a 524td – and was available only with the five-speed overdrive transmission. It was also a fairly basic entry-level model on which items such as ABS cost extra. Sales were never very strong, and the car was not built in South Africa and not offered for sale in the USA. While many experienced BMW owners felt that the 518i did not offer the full BMW driving experience, it nevertheless brought the general excellence of the E34 range to a new group of buyers, many of whom probably moved on to the bigger-engined models at a later date.

FOUR-VALVE ENGINES IN THE 520i AND 525i, 1990

The E34s had been on sale for just over two years when BMW made some major adjustments to the range. In May 1990, the slow-selling 530i went out of production and was not directly replaced. However, its 188bhp were eclipsed by the 192bhp in the revised 525i which arrived the same month. At the same time, the 520i was re-engined with a new 150bhp power unit.

The new 520i and 525i engines were actually four-valve developments of the

> **Production of the E34 5 Series**
>
> The E34 5 Series was produced in far greater numbers than its two predecessor ranges, the E12 and E28 models. The millionth example left the production lines at Dingolfing on 16 September 1993, and by the middle of 1994 a further 150,000 had been made and production was running at a rate of 730 cars a day. Of this daily total, a maximum of 12 would be M5s, hand-built by the 100 specialist staff of the Motorsport division at Garching.

earlier engines, that remained in production until July 1990. Now equipped with twin overhead camshafts to operate the more complex valve train, they had also been redesignated M50 types. Four-valve engines had of course first been seen in BMWs prepared by the Motorsport division; the works racing coupés had used them in the mid-1970s, and then they had gradually filtered down through the M1 in 1978 to the M 635CSi in 1983 and of course to the E28 M5 in 1985 and the E30 M3 a year later. With that much experience of four-valve engines behind them, the BMW engineers were prepared to release these inherently more complex engines for use in everyday family saloons like the mid-range E34s. And so confident were they of the durability of these new engines that they reduced their servicing requirements as compared to their predecessors.

The new 24-valve M50 sixes certainly offered more power than the M20s with identical swept volumes, but their new configuration did not bring only gains. They were not quite as refined as the earlier engines, and the main performance advantage they brought was better high-speed acceleration thanks to a lot more

Third Generation: The E34 Models, 1988–96

The Touring models had an integrated design and did not look like modified saloons. The side marker lights are the only clue that this is a US-market version.

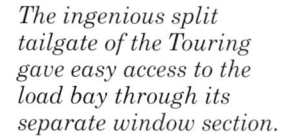

The ingenious split tailgate of the Touring gave easy access to the load bay through its separate window section.

peak torque at the same high engine revolutions. For the long German Autobahns with their absence of speed limits, the cars brought some worthwhile advantages. For other countries such as Britain (with a 70mph speed limit) and the USA (with a 55mph speed limit in most states), the improved high-speed acceleration was fairly academic. Both the new engines demanded Super Unleaded petrol instead of the cheaper Normal Unleaded that had contented their

predecessors, and the 520i proved to be thirstier in everyday use although the new 525i was slightly more economical on fuel than the old two-valve car.

NEW TRANSMISSIONS AND THE 1991 MODEL-YEAR CHANGES

The new four-valve engines were introduced some months ahead of the 1991

Third Generation: The E34 Models, 1988–96

The well-balanced lines of the E34 were equally successful from the rear. Note the stepped shape of the tail light clusters.

model-year changes announced in August 1990, just before that year's Frankfurt Show. That month saw a number of equipment upgrades for the new 1991 model-year E34 range, as exhaust catalysts were standardized right across the range for all territories; air conditioning was also made standard for most markets, and an air micro-filter that excluded pollen and dust from the passenger compartment was made optional. There were also some important drivetrain changes for some models.

Most far-reaching was the arrival of two new gearboxes for the 520i and 525i models, that had been available for only a matter of months with their four-valve engines harnessed to the older gearboxes. In place of the overdrive four-speed automatic was a new ZF overdrive five-speed transmission with torque converter lock-up in both fourth and fifth gears and with closer ratios in the lower gears for better acceleration. Meanwhile, a new and lighter five-speed direct-top manual gearbox replaced the older overdrive type, once again giving a better spread of ratios in the intermediates to improve in-gear acceleration. With both new transmissions, the axle ratios were raised in order to maintain fuel economy and high maximum speeds. In a complementary move, the overall gearing of the 535i was lowered to give it a more convincing acceleration advantage over the new smaller-engined cars.

THE TOURING MODELS, 1991

> While it might not be the most voluminous estate car on the market, the 5 Series Touring is one of the most appealing. It inherits most of the finesse of the saloon, and performs its haulage tasks with enough competence to satisfy 90 percent of potential owners.
>
> *Car*, May 1992

The E34 revisions introduced in 1990 were only a taster of what was to come, however, because BMW had been preparing some further radical improvements for the range. They arrived during 1991, in the shape of a new estate bodyshell, a bigger and more powerful turbodiesel engine, and

107

Third Generation: The E34 Models, 1988–96

Dark colours suited the Touring as well as light ones – a sure sign of sound styling. This is another US-market car.

The Touring models shared their high performance and handling dynamics with the saloons. This one was used for fast-response work as the course doctor's car at Britain's Thruxton racing circuit.

Boom and Bust

Careful marketing allowed BMW to exploit to the full the economic boom in Western countries during the late 1980s. There was enough money in the Western economies for large numbers of people to afford the high showroom prices of German cars, and BMW trod a careful line between catering for demand and maintaining exclusivity through high-priced optional extras. The year 1990 proved to be the best-ever in BMW's 75-year history. Turnover and profit achieved record levels and profit after taxation increased 25 per cent.

However, the boom was just about to end. By mid-1991, the car market was feeling the effects of recession, and BMW (GB) announced that it planned to save half a million pounds by not attending the London Motorfair (that alternated at the time with the bi-annual Motor Show held in Birmingham).

The 1994 Family Touring prototype was an attempt to design a car interior specially to suit young families. Some of its ideas later entered production in the E39 Touring models.

a four-wheel drive option for the 525i. At the same time, ABS was standardised right across the E34 range.

The most far-reaching of these changes was undoubtedly the arrival of the estate bodyshell that had been an open secret for several months before its public appearance at the Frankfurt Show in September 1991. Marketed under the Touring name that had been used on the hatchback 2002 Touring between 1971 and 1974, this was the first-ever factory-built estate car from BMW, and its introduction had been provoked by the success of medium-sized estate cars from Mercedes-Benz and Volvo. Thus, it was faced with the difficult task of stealing sales from two well-established and highly-respected products.

Perhaps the most remarkable characteristic of the new Touring shell was that its rear section was so well integrated with the original saloon body. Like the Mercedes, the two bodies had clearly been designed in tandem so that the estate did not look like a cobbled-together aftermarket conversion; the window frames of the rear side doors, that so often give away the saloon origins of an estate, had even been modified to suit the new roof

> ### E34 Security
>
> BMW put a lot of effort into car security for the E34 5 Series cars. From the beginning, all models had deadlocks as standard. On cars fitted with an on-board computer, it was possible to use this as an engine immobilizer; the engine could only be re-started when the correct code had been entered into the computer.
>
> In September 1991, the VIN was stamped on a plate that could be read through the windscreen (thus enabling Police to compare it with records and detect when a car had been fitted with false number plates). Then from September 1992, an engine immobilizer was fitted across the range, this time actuated by the deadlocks on the doors.
>
> Two further changes were made in January 1995, when all door and boot locks were fitted with freewheeling lock cylinders, and a rolling-code engine immobiliser system known as EWS was introduced. This depended on a transponder-type ignition key with a chip moulded into it that was read and modified by a processor on the car. This processor selected a new random code each time the key was inserted in the ignition, so that it was impossible to copy or overcome the system. New keys were of course obtainable only from BMW.

line. Extra cost options included two separate electrically-operated tilt-and-slide sunroofs, one for the front passengers and one for those in the rear, and roof rails which made a roof rack when fitted with their removable crossbars.

The sloping tailgate hinged from the roof in the conventional fashion, but its glass section could be opened independently to allow small items to be put into the load area. Neat details included electrically assisted catches for

Third Generation: The E34 Models, 1988–96

The turbodiesel 525tds was a strong-selling addition to the range, offering the expected BMW levels of refinement, excellent performance, and reduced fuel costs.

both window and main tailgate, a wiper unit that detached from its motor when the glass section was lifted, and a washer jet that appeared out of the panelwork only when required. The load compartment floor was fitted with four retractable luggage securing points, and there was an additional storage compartment concealed beneath it. The rear seat had a one-third/two-thirds split and could be folded forwards with one hand to give extra length in the load area. It came as standard with two head restraints and optionally with three, and none of these had to be removed before the set could be fully folded. The finishing touch was a rollaway blind to conceal the contents of the load area, and of course a boot liner and dog guard were extra-cost options.

The Touring body was not made available with all the engine variants of the E34 range at first, but could only be had as a 520i, 525i, 525iX or 525tds; BMW's view was that it would have no appeal to buyers at the two extremes of the range, and so there was no 1.8-litre version, no 3.5-litre version and – initially at least – no M5 version. Specification levels generally matched those of the equivalent saloons, but increased weight meant that the 520i took on ventilated front brake discs. Right-hand drive variants lagged some months behind left-hand drive models, and the Touring did not go on sale in Britain until March 1992. At

---

Sport and SE Models

There was no such thing as a single worldwide specification for any of the E34 models. Importers in each country where BMW sold its cars made up a specification to suit local market conditions, adding their own cocktail of extras to a 'base' specification that was common to all markets.

Britain was BMW's best European market outside its native Germany, and the company therefore made special efforts to meet customer requirements there, as interpreted by BMW (GB). So it was that the British market was offered E34 models known as Sport and SE types. These cars did not carry special badging and were simply 'standard' variants put together out of BMW's extensive options list.

Typical of the SE specification (in this case relating to a 518i SE in mid-1990) was the addition of cross-spoke alloy wheels, heated door locks, an electric sunroof and a leather-bound steering wheel. Typical of the Sport specification (for a 1991 525i Sport) were sports suspension, sports seats, wide cross-spoke alloy wheels, a limited-slip differential and a bodykit of aerodynamic addenda. From May 1994, the UK-market 525i Sport took on a number of items from the M5 specification: cloth-and-leather combination seats, Bird's-Eye Maple wood trim, a black headlining and the special Avus Blue and Daytona Violet paint options. SE and Sport models were available with either manual or automatic transmission.

Third Generation: The E34 Models, 1988–96

that stage, it was announced with just two variants – 520i and 525i. The 525iX Touring followed a few months later.

In fact, the success of the E34 Touring seems to have caught BMW rather by surprise, and demand built up faster than supply. Over the next few years, it became both an exclusive and a highly prized possession – and, to the annoyance of its owners, inevitably became a common target for car thieves.

THE 525iX, 1991

> It drives, to all intents and purposes, just like a rear-wheel-drive 525i – which means it's quiet, comfortable, smooth, stable and quick.
>
> *BMW Car*, January 1995

Four-wheel drive on a road car suddenly became fashionable after the success of the 1980 Audi Quattro coupé that had been developed as a rally car. Mercedes-Benz introduced its 'intelligent' 4-Matic system as a traction control aid at the 1985 Frankfurt Show, and Ford announced 4x4 options for the Sierra and Granada Scorpio at the same time, but BMW held

The Recyclable BMW

In the early 1990s, BMW put a lot of effort into environmental protection and recycling. When introduced in 1991, the Touring models were claimed to be over 80 per cent recyclable. In addition to the cleaner exhausts resulting from the catalytic converters demanded by law in many countries, BMW had also looked into the question of natural resources used in the production process. Thus the company was able to claim in 1992 that it used only 2 cubic metres of water in manufacturing each car, as compared to a German car industry average of 13 cubic metres per car.

BMW take great care over the presentation of their engines. This is the 2.5-litre intercoooled turbodiesel six under the bonnet of a 525tds.

Third Generation: The E34 Models, 1988–96

Engines in the E34 Models

As with the previous 5 Series ranges, the engines in the E34 models were drawn from several different BMW engine families.

Four-cylinder
The only four-cylinder engine in the E34 range was the 1796cc M43 type in the 518i. This started life with its overhead camshaft driven by a toothed belt, but after March 1993 was re-engineered with a single roller chain drive, hydraulic tappets and reduced internal friction. The engine was new in 1987, when it was first seen in the E30 318i. In the 518i, its block was canted over to the right by 20 degrees.

Small sixes
There were two varieties of the M20 small-block six in the E34 models, a 1991cc edition in the 520i and a 2494cc edition in the 525i. The M20 was an iron-block, alloy-head engine; it had started life with a single chain-driven overhead camshaft, but in the E34 models was fitted with a toothed belt camshaft drive. It was canted to the right by 20 degrees.

The M20 was developed further to become an M50 in 1990, when the same basic engine was given a new cylinder head with twin overhead camshafts and four valves per cylinder. The four-valve engines had the same swept volumes as the M20s and went into the 520i, 525i and 525iX models. They also had a more advanced version of the Bosch DME (Digital Motor Electronics) engine management system and their camshafts were driven by roller chains.

From 1992, the 2-litre and 2.5-litre engines were further modified by the addition of VANOS variable valve timing.

The M51 turbodiesels in the 525td and 525tds models were related to these M50 small sixes, but did not have the four-valve cylinder heads. Instead, they had single overhead camshafts and indirect injection, with DDE II (Digital Diesel Electronics) engine management systems. Both had the same 2498cc capacity, the engine in the tds models being equipped additionally with an air-to-air intercooler. The 2443cc M21 turbodiesel in the early 524td represented an earlier stage in the development of the compression-ignition engines, and was based on the earlier M60 small-block six

Big sixes
The M30 big-block six-cylinder engines were by now elderly, but two different varieties saw service in the E34 range. They powered the 530i (2985cc) and 535i (3430cc) models. Related to them were the two M88 Motorsport engines in the M5 (3535cc and later 3795cc). All had iron blocks and alloy heads; there were single overhead camshafts in the 530i and 535i engines, but the Motorsport-developed M5 engines both had twin overhead camshafts and four-valve cylinder heads. In all cases, the camshaft drive was by duplex chain. All four engines were installed with their blocks canted at 30 degrees to the right.

V8s
The M60 V8 engines came with a 2997cc capacity for the later 530i and a 3982cc capacity for the 540i. Both were introduced in the big 7 Series saloons at the same time as they apeared in the E34s, and the larger engine was also used in the 8 Series coupés from 1993. They were all-alloy engines with chain-driven twin overhead camshafts, hydraulic tappets and four valves per cylinder.

The turbodiesel engine shared much of its design with the small-block six-cylinder petrol engines.

back until the beginning of the 1990s. Like the Mercedes 4-Matic, the system that BMW introduced in 1991 on both the 3 Series and E34 5 Series cars was designed to add to the safety of the car in everyday driving situations rather than to compensate for any deficiencies in the basic handling characteristics of the chassis; in particular, according to BMW, it had been developed for winter conditions in the Bavarian and Austrian Alps. However, it was costly and, like the Mercedes and Ford systems, never sold well. So BMW discontinued it and it did not reappear on the fourth-generation E39 cars when these were introduced in 1996.

In the E34 range, the only model available with four-wheel drive was a variant of the 525i called the 525iX. Like the 325iX introduced at the same time, it had a permanently engaged four-wheel drive system that used sensors to adjust the torque distribution to all four wheels in order to meet traction demands. In normal driving, 36 per cent of the torque went to the front wheels and 64 per cent to the rear. However, sensors at each wheel constantly monitored traction, while others monitored road speed and driver input such as throttle and braking. When wheelspin was detected (that is, one or more wheels had lost traction and therefore started to spin faster than the others), the computer 'brain' instantly redistributed the torque to compensate. This was achieved through an electromagnetic centre differential (that could shift all the torque to the front or the rear pair of wheels if necessary) and an electro-hydraulic differential at the rear. These adjustments made by the system were undetectable to the driver, although a dashboard light would flash to warn him that the system was in action and that there may have been surface traction problems. In addition, the centre differential disconnected the 4x4 system automatically as soon as the brakes were applied to allow the ABS to function correctly. As so often happened with BMW's new technology, right-hand drive cars were not equipped with the 4x4 system until some months after it had entered production for left-hand drive markets – in this case, in June 1992 – and in Britain the 525iX was always available only to special order.

THE 525tds, 1991

The third introduction at the Frankfurt Show in 1991 was a superb new diesel engine. The old 524td model ceased production and the larger capacity of the new engine allowed BMW to call the new

Third Generation: The E34 Models, 1988–96

BMW E34 5 Series models, 1988–96

All models shared the same basic architecture of a unitary four/five-seater bodyshell with front and rear crumple zones, with a front-mounted engine driving the rear wheels. The 525iX only had permanent four-wheel drive.

518i SALOON (1989–96) AND TOURING (1993–96)

Engine:
Cylinders	1796cc four-cylinder
Bore and stroke	84mm x 81mm with overhead camshaft
Compression ratio	8.8:1
Carburettor	Bosch Motronic fuel injection exhaust with catalytic converter and Lambda sensor
Max power	113bhp at 5500rpm
Max torque	117 lb/ft at 4250rpm.

Transmission:
Gearbox	Five-speed overdrive manual
Top	3.72:1
4th	2.04:1
3rd	1.34:1
2nd	1.00:1
1st	0.80:1
1989-1992	Five-speed close-ratio manual gearbox and optional four-speed overdrive automatic from March 1993.
Gear ratios (manual)	
Top	5.10:1
4th	2.77:1
3rd	1.72:1
2nd	1.22:1
1st	1.00:1
Gear ratios (automatic)	
	2.40:1
	1.47:1
	1.00:1
	0.72:1
Axle ratio	4.27:1 (1989–93)
	3.46:1 (manual, from March 1993)
	4.45:1 (automatic). Automatic not available on Touring models.

Suspension and steering
	Independent front suspension, with MacPherson struts and anti-roll bar; semi-trailing arm rear suspension with coil-sprung struts attached to the hub carriers.
Steering	ZF Gemmer worm and roller steering with servo assistance.
Tyres	195/65 HR 15 tyres
wheels	6J x 15 steel wheels (saloon)
	205/65 HR 15 tyres

	7J x 15 steel wheels (Touring).
Brakes	Servo-assisted all-disc brakes with dual hydraulic circuit ABS standard from 1993.
Dimensions	
Overall length	4,720mm (185.8in)
Overall width	1,751mm (68.9in)
Overall height	1,412mm (55.6in) for saloon 1,417mm (55.8in) for Touring
Wheelbase	2,761mm (108.7in)
Front track	1,470mm (57.9in)
Rear track	1,495mm (58.8in).
Unladen weight	1400kg (3086 lb) for saloon or 1485kg (3274 lb) for Touring.

520i SALOON (1988–96) AND TOURING (1991–96)

As for 518, except:
Engine

Cylinders	1991cc six-cylinder
Bore and stroke	80mm x 66mm with overhead camshaft,
Compression ratio	8.8:1
Carburettor	Bosch Motronic fuel injection exhaust with catalytic converter and Lambda sensor
Max power	129bhp at 6000rpm
Max torque	118.5 lb/ft at 4300rpm (1988–90).
From May 1990	twin overhead camshafts and four-valve cylinder head
Compression ratio	10.5-11.0
Max power	150bhp at 5900rpm
Max torque	137 lb/ft at 4300rpm.
From September 1992	VANOS variable valve timing on the inlet camshaft.

Transmission

Gearbox	Five-speed overdrive manual Optional four-speed overdrive automatic 2.48:1 1.48:1 1.00:1 0.73:1
Axle ratio	4.45:1 (1988–90).
From May to July 1990	Five-speed overdrive manual
Top	3.83:1
4th	2.20:1
3rd	1.40:1
2nd	1.00:1
1st	0.81:1
Optional four-speed overdrive automatic	
	2.40:1

	1.47:1
	1.00:1
	0.72:1
Axle ratio	4.27:1 (manual)
	4.55:1 (automatic).
From August 1990	Five-speed close-ratio manual
Top	4.23:1
4th	2.52:1
3rd	1.66:1
2nd	1.22:1
1st	1.00:1
Optional five-speed overdrive automatic	
	1.366:1
	2.00:1
	1.41:1
	1.00:1
	0.74:1
Axle ratios	3.46:1 (manual)
	3.64:1 (automatic).
Tyres	195/65 HR 15
(from May 1990)	195/65 VR 15
Wheels	6J x 15 steel (saloon)
	205/65 HR 15 tyres
	7J x 15 steel wheels (Touring).
Brakes	ABS standard from September 1991.
Dimensions	
Unladen weight	1480kg (3262 lb) for manual saloon
	1510kg (3329 lb) for automatic (1988-1990).
From May 1990	1475kg (3252 lb) for manual saloon.
Touring models	1550kg (3417 lb) for manual or 1585kg (3494 lb) for automatic.

524td SALOON (1988–91)

As for 518, except:
Engine:
Cylinders	2443cc six-cylinder
Bore and stroke	80mm x 81mm indirect-injection turbocharged diesel with overhead camshaft
Compression ratio	22:1
Carburettor	Garrett T 03 turbocharger and Bosch DDE injection
Max power	115bhp at 4800rpm
Max torque	159 lb/ft at 2400rpm

Oxidation catalyst optional from spring 1990.

Transmission:
Gearbox	Five-speed overdrive manual

Third Generation: The E34 Models, 1988–96

Top	4.35:1
4th	2.33:1
3rd	1.39:1
2nd	1.00:1
1st	0.81:1
Optional four-speed overdrive automatic	
	2.73:1
	1.56:1
	1.00:1
	0.73:1
Axle ratios	3.25:1 (manual)
	3.46:1 (automatic).
Tyres	195/65 HR 15
	195/65 VR 15 tyres
Wheels	6J x 15 steel

Brakes   ABS standard.

Dimensions
Unladen weight   1480kg (3263 lb) for manual
                              1510kg (3329 lb) for automatic.

525i SALOON (1988–96), 525i TOURING (1990–96), 525ix SALOON AND TOURING (1991–96)

As for 518, except:
Engine

Cylinders	2494cc six-cylinder
Bore and stroke	84mm x 75mm with overhead camshaft
Compression ratio	8.8:1
Carburettor	Bosch Motronic fuel injection
	exhaust with catalytic converter and Lambda sensor
Max power	170bhp at 5800rpm
Max torque	160 lb/ft at 4300rpm.
From May 1990	twin overhead camshafts and four-valve cylinder head
Compression ratio	10.0–10.5:1
Max power	192bhp at 5900rpm
Max torque	180 lb/ft at 4500rpm.
From September 1992	VANOS variable valve timing on the inlet camshaft.

Transmission

Gearbox	Five-speed overdrive manual
Top	3.83:1
4th	2.20:1
3rd	1.40:1
2nd	1.00:1
1st	0.80:1

Third Generation: The E34 Models, 1988–96

Optional four-speed overdrive automatic
                    2.48:1
                    1.48:1
                    1.00:1
                    0.73:1
Axle ratio   3.73:1 (manual)
                    3.91:1 (automatic), 1988–90.
From May to July 1990   Five-speed overdrive manual gearbox or optional four-speed overdrive automatic (all ratios as for 520i)
Axle ratio   3.73:1 (manual)
                    4.10:1 (automatic).
From August 1990   Five-speed close-ratio manual gearbox
Top   4.23:1
4th   2.49:1
3rd   1.66:1
2nd   1.24:1
1st   1.00:1
Optional five-speed overdrive automatic (ratios as for 520i)
Axle ratio   3.23:1.
525iX models   As for post-August 1990 525i, but with permanent four-wheel drive and 3.38:1 axle ratio.

Suspension and steering
Steering (525iX only)   Variable-ratio rack-and-pinion (??).
Tyres: 195/65 VR 15
Wheels   6.50J x 15 steel
From May 1990   6J or 6.50J steel wheels.
525i Touring   225/60 VR 15 tyres
                    7J x 15 steel wheels.
525iX   225/55 HR 16 tyres
                    7.50J x 16 steel wheels.
525iX Touring   225/55 VR 16 tyres.

Brakes   ABS optional to August 1991 and standard from September.

Dimensions
Front track   1479mm (58.2in) from May 1990
                    1468mm (57.8in) on 525iX models.
Unladen weight   1530kg (3373 lb) for manual
                    1560kg (3439 lb) for automatic, 1988-1990.
From May 1990   1525kg (3362 lb) for manual saloon
                    1595kg (3516 lb) for manual Touring
                    1630kg (3593 lb) for automatic.
525iX   1610kg (3549 lb) for manual
                    1640kg (3615 lb) for automatic saloon
                    1670kg (3682 lb) for manual Touring
                    1705kg (3759 lb) for automatic.

525td SALOON AND TOURING (1993–96)

As for 518, except:
Engine
Cylinders   2498cc six-cylinder
Bore and stroke   80mm x 82.8mm indirect-injection turbocharged diesel
                         with overhead camshaft
Compression ratio   22:1
Carburettor   Garrett T 03 turbocharger and Bosch DDE injection
                         oxidation catalyst
Max power   115bhp at 4800rpm
Max torque   160 lb/ft at 1900rpm.

Transmission
Gearbox   Five-speed close-ratio manual
Top   5.09:1
4th   2.80:1
3rd   1.76:1
2nd   1.25:1
1st   1.00:1)
Optional four-speed overdrive automatic
                         2.86:1
                         1.62:1
                         1.00:1
                         0.72:1
Axle ratio   2.65:1 (manual saloon)
                         2.79:1 (manual Touring)
                         3.26:1 (automatic).
Brakes   ABS standard.

Dimensions
Unladen weight   1485kg (3274 lb) for manual saloon
                         1520kg (3351 lb) for automatic
                         1565kg (3450 lb) for manual Touring
                         1600kg (3527 lb) for automatic.

525tds SALOON AND TOURING (1991–96)

As for 525td, except:
Engine
Cylinders   Fitted with air-to-air intercooler
Max power   143bhp at 4800rpm
Max torque   188 lb/ft at 2200rpm.

Transmission
Gearbox   Optional five-speed overdrive automatic
                         3.67:1

	2.00:1
	1.41:1
	1.00:1
	0.74:1
Axle ratio (manual and automatic, saloon and Touring)	
	2.65:1.
Unladen weight	1500kg (3307 lb) for manual saloon
	1535kg (3384 lb) for automatic
	1580kg (3483 lb) for manual Touring
	1615kg (3560 lb) for automatic.

530i (1988–90)

As for 518, except:
Engine

Cylinders	2986cc six-cylinder
Bore and stroke	89mm x 80mm with overhead camshaft
Compression ratio	9.0:1
Carburettor	Bosch Motronic fuel injection
	exhaust with catalytic converter and Lambda sensor
Max power	188bhp at 5800rpm
Max torque	188lb/ft at 4000rpm.

Transmission:

Gearbox	Five-speed overdrive manual
Top	3.83:1
4th	2.20:1
3rd	1.40:1
2nd	1.00:1
1st	0.81:1
Optional four-speed overdrive automatic	
	2.48:1
	1.48:1
	1.00:1
	0.73:1
Axle ratio	3.64:1 (manual)
	3.73:1 (automatic).
Tyres	205/65 VR 15
Wheels	6.5J x 15 steel

Brakes   ABS standard.

Dimensions
Unladen weight: 1590kg (3505 lb) for manual or 1610kg (3549 lb) for automatic.

530i SALOON AND TOURING (1992–96)

As for earlier 530i, except:
Engine
Cylinders	2997cc V8-cylinder
Bore and stroke	84mm x 67.6mm with twin overhead camshafts and four-valve cylinder heads
Compression ratio	10.5:1
Carburettor	Bosch Motronic fuel injection exhaust with catalytic converter and Lambda sensor
Max power	218bhp at 5800rpm
Max torque	209 lb/ft at 4500rpm.

Transmission
Gearbox	Five-speed close-ratio manual gearbox
Top	4.20:1
4th	2.49:1
3rd	1.66:1
2nd	1.24:1
1st	1.00:1
Optional five-speed overdrive automatic	
	3.67:1
	2.00:1
	1.41:1
	1.00:1
	0.74:1
Axle ratio	3.08:1 (manual saloon)
	3.15:1 (automatic saloon)
	3.23:1 (all Touring models).

Suspension and steering
Suspension	Rear self-levelling on Touring models.
Tyres	225/60 ZR 15
Wheels	7J x 15 alloy

Dimensions:
Overall height	1412mm (55.8in) for Touring models.
Unladen weight	1610kg (3549 lb) for manual
	1630kg (3593 lb) for automatic.

535i (1988–91)

As for 518, except:
Engine:
Cylinders	3430cc six-cylinder
Bore and stroke	92mm x 86mm with overhead camshaft
Compression ratio	9.0:1

Third Generation: The E34 Models, 1988–96

Carburettor	Bosch Motronic fuel injection
	exhaust with catalytic converter and Lambda sensor
Max power	211bhp at 5700rpm
Max torque	220 lb/ft at 4000rpm.

Transmission:
Gearbox — Five-speed overdrive manual gearbox or optional four-speed overdrive automatic (all ratios as for six-cylinder 530i)
Axle ratio — 3.45:1 (manual, to July 1990)
3.64:1 (manual, from August 1990)
3.46:1 (automatic, to July 1990)
3.91:1 (automatic, from August 1990).

Suspension and steering
Tyres — 225/60 VR 15 or 225/60 ZR 15
Wheels — 7J x 15 alloy wheels.

Brakes — ABS standard.

Dimensions
Unladen weight: 1620kg (3571 lb) for manual or 1640kg (3615 lb) for automatic.

540i (1992–96)
As for V8-engined 530i, except:
Engine
Cylinders — 3982cc V8-cylinder
Bore and stroke — 89mm x 80mm swith twin overhead camshafts and four-valve cylinder heads
Compression ratio — 10.0:1
Carburettor — Bosch Motronic fuel injection
exhaust with catalytic converter and Lambda sensor;
Max power — 286bhp at 5800rpm
Max torque — 289 lb/ft at 4500rpm.

Transmission:
Gearbox — Five-speed overdrive automatic only
3.55:1
2.24:1
1.54:1
1.00:1
0.79:1)
Axle ratio — 2.93:1
Dimensions
Unladen weight: 1650kg (3637 lb).

Third Generation: The E34 Models, 1988–96

ones by the 525tds designation. That 's' also had a special significance, because it reflected BMW's confidence in pitching the new model at a higher level in the market than the old. Not every market took 525tds, however. None were ever made in South Africa, and it was not until 1993 that right-hand drive examples were made for Britain.

The new engine was a development of the older one, but it was very different in many respects and carried the new M51 designation. Its larger capacity of 2.5 litres (2498cc) had been achieved by lengthening the stroke, and that was one factor behind its much higher torque at lower crankshaft speeds. However, the engine also had lighter pistons with reinforced gudgeon pins, and its oil pump was now driven directly from the crankshaft instead of through an intermediate shaft and an angle drive. The engine wiring harness was now protected in a cable shaft, and the fuel was warmed up in the filter head itself without the need for external components. As before, there was sound-deadening encapulation in the engine bay, but this time with temperature-controlled flaps that closed to speed the warm-up process.

The M51 turbodiesel also benefited from two major improvements. The first of these was an air-to-air intercooler that lowered the temperature of the charge air to make it more dense and give better combustion. The second was the latest-generation Bosch DDE II engine management system that gave more accurate fuel metering to promote more efficient combustion. This in turn made for smoother running, reduced noise, lower consumption and lower emissions. In addition, it did not suffer from cyclical fluctuations and so made the engine run more smoothly at all times.

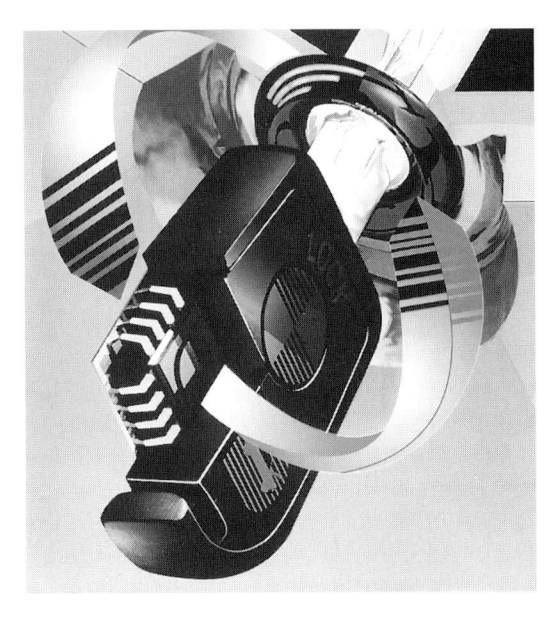

This is the EWS rolling-code immobiliser system. A transponder in the ignition key 'talked' to the car and an on-board processor reset the code each time. The system offered such a huge variety of code combinations that it was impossible to beat.

THE V8 ENGINES, 1992

The top-model 535i, hastily revised in 1990 to keep it saleable alongside the new four-valve 525i, disappeared at the end of the 1992 season to be replaced by two new models. One took the dormant 530i designation and the other created a new 540i. Both had brand-new four-valve 90-degree V8 engines of all-alloy construction – the first BMW production V8s for 27 years.

The new M60 V8s had actually been announced in the E32 7 Series cars at the Geneva Show in March 1992, but it was clear from the beginning that they had been designed to fit under the bonnets of the entire BMW range. (Aftermarket conversions have shown that they will even

Third Generation: The E34 Models, 1988–96

fit into the E36 3 Series cars, although BMW have never offered such a model.) So no-one was very surprised when they were announced for the 5 Series in July 1992.

Work had started on these engines back in 1984, but in the beginning it was by no means clear whether the V8 design would win the day over a 4-litre all-alloy straight-six which was under development at the same time. By 1986, however, BMW had settled on the V8s. They were an all-new design, sharing almost no common elements with existing BMW production engines and – surprisingly – they shared few common components with one another. Bore and stroke were different, so that the engines could not share either cylinder blocks or crankshafts. In fact, the 218bhp 3-litre engine had a cast crankshaft, while the 286bhp 4-litre had a forged crankshaft.

Weight-saving had been a fundamental design aim in creating the new engines, and in full trim they weighed-in at an extremely light 447lb (203kg). Computer-aided design had enabled the weight of the cylinder block to be kept down to an astonishing 55lb (25kg), while the cylinder heads each weighed a mere 65lb (30kg). Weight had also been saved in smaller areas: thus, the powder-metal con-rods were 18 per cent lighter than forged steel equivalents, while the cylinder head covers were made of lightweight magnesium and the airbox was made of plastic that, in common with BMW's latest policies – was fully recyclable.

The V8s had a total of four overhead camshafts – two per cylinder bank – to operate their valves via hydraulic tappets. The crankshaft pulley drove the inlet camshafts directly by a long duplex chain, and the exhaust camshafts were driven by short secondary chains. Each chain also had guide rails and tensioners. Engine management was by the latest Bosch DME 3.3 system, that operated the multi-point fuel injection and the direct ignition with eight individual coils, each one positioned alongside its corresponding spark plug. And, of course BMW had paid careful attention to the appearance of the engines. Once the bonnet was opened, most of what was visible was moulded cosmetic covers; the engine in all its complexity was buried beneath them.

Performance from these two new engines was exemplary. In the time-honoured BMW way, both delivered their power smoothly and quietly, with a discreet but nonetheless distinctly sporting exhaust note. The 3-litre needed to be revved quite hard to give of its best, but it offered both quicker acceleration through the gears and a higher top speed than the superseded 535i. As for the 540i, which was available only in saloon form and only with the five-speed automatic transmission, it provided stunning performance right up to its 155mph (250kph) limited top speed.

THE VANOS ENGINES, 1992

September 1992 also brought a very important change to the two small six-cylinder engines, which were modified by the addition of variable camshaft timing. This was known as VANOS, a name which came from the German VAriable NOckenwellen Steuerung (literally: variable camshaft control). The VANOS engines were easily identified from their predecessors by a glance under the bonnet, because they had a boss at the forward end of the camshaft cover instead of a flat front face.

The system operated on the inlet camshaft only, and was controlled by the DME engine management system. The inlet camshaft was rotated slightly relative

Third Generation: The E34 Models, 1988–96

to its driving chain by moving forwards or backwards in a helical gear within a housing at the front of the engine, and this movement was controlled by engine oil released to one or the other side of a control piston. Forward movement had the effect of retarding the timing, and rearward movement advanced it. The whole process was completely smooth and stepless, and gave the 520i and 525i engines more valve overlap for high-speed work and less overlap for low-speed running and idling.

With the gas flow into the cylinders thus regulated to nearly ideal levels for all conditions, the engine management system was able to adjust ignition and injection in order to provide better fuel consumption and stronger low-speed torque. Thus, at 3500rpm, the 520i engine with VANOS gave 7 per cent more torque than the non-VANOS four-valve engine. The 525i engine saw a 10 per cent improvement, and in addition its torque curve peaked at 1200rpm below that of the old engine. The results were much improved driveability at all speeds.

To go with the new V8 and VANOS engines for the 1993 model-year, BMW also announced a refinement to their automatic transmissions. Known as AGS (Adaptive Gear Shift), this was an electronic control system which adapted its shift points automatically to suit the driver's style. In addition, it offered full manual control

As so often on the 5 Series cars, wheel design made an important difference. Compare the cross-spoke alloys of this US-market Touring with the simpler radial-spoke design pictured on p.96.

Third Generation: The E34 Models, 1988–96

and a choice of four shift programmes: Economy, Normal, Sports and Winter.

THE 525td AND REVISED 518i, 1993

The 1994 model-year brought two new models as far as the German market was concerned, although one of those was the 518i that had been available for export since 1989. The 518i announced in autumn 1993 was not quite the same car as before, however, and the new version replaced the old 518i in all markets that took the 1.8-litre E34. The revised car had the new five-speed direct-top manual gearbox with taller overall gearing and could be ordered optionally with the old overdrive four-speed automatic. In addition, a Touring version was made available for the first time, although only with the manual transmission.

Introduced at the same time was a lower-powered turbodiesel model, badged as a 525td and available in both saloon and Touring forms. The lower power had been achieved by the simple expedient of removing the engine's intercooler, and the cheaper showroom price was a result of this and of removing some items of equipment. The non-intercooled turbodiesel was equipped with an oxidation catalyst, but nevertheless boasted the same power output as the 2.4-litre engine that had gone out of production two years earlier so that the 525td had the same performance as the older 524td. Transmissions were the direct-top five-speed manual and the overdrive four-speed automatic; like the 518i, the engine's characteristics made it unsuitable for use with the latest overdrive five-speed automatic.

There were other changes for the 1994 model-year. Smaller 'Euro' airbags were fitted to all models except the 525iX and the Touring variants, that retained the larger US airbags. These were, of course, still fitted to all US-market variants of the E34 range. The 530i and 540i both took on the latest AGS five-speed automatic transmissions, and the pollen filter was standardized across the range. Dechroming, first seen as an option on the 7 Series cars and the M5, was also made optional on the E34s.

A FACELIFT, A NEW GEARBOX AND THE 540i TOURING, 1994

By this time, the E34 range was in its twilight years and BMW was preparing to replace it with the new E39 cars. However, changes continued to appear. At the Geneva Show in March 1994, a new and broader radiator grille replaced the original, and alloy wheels and burr walnut interior wood trim were standardised across the range. A 540i Touring was added to the line-up, and passenger's side airbags were made optional on all models. The 518i had engine modifications which gave better torque and lower fuel consumption, and the two V8 models were equipped with an automatic speed hold and leather trim as standard. On the 540i, a manual gearbox option was offered for the first time, in the shape of the six-speed type that was made available for the M5 at the same time.

THE FINAL YEARS, 1995–96

Detail changes continued to appear, making clear that BMW policy was still one of continuous improvement. The 1995-

Third Generation: The E34 Models, 1988–96

model cars built after August 1994 had a seat sensor linked to the passenger's side airbag when fitted, so that the airbag would not be activated if the seat was unoccupied. In addition, a 'situation sensor' prevented airbag deployment in low-speed collisions. In January 1995 there were some important anti-theft upgrades, and in June a retro-fit air-conditioning kit was made available for right-hand drive cars. It always had been possible to add air conditioning to a car built without it, of course, but the need to change major items like the alternator, radiator and coil springs had made the exercise prohibitively expensive. With the new kit, costs were now contained within reasonable bounds.

BMW issued the first pre-launch publicity pictures and information about the new E39 models in May 1995. The new fourth-generation 5 Series cars were announced at the Frankfurt Show that September, but the E34s remained on sale well into 1996 until all the new models had been prepared for all markets. In fact, there were distinctive 1996 model-year changes for the E34s, all of which gained a high-level central brake light while the in-car entertainment system of the 530i and 540i Touring models was also upgraded. The Touring models remained in production after the saloons to cover the gap before the new E39 Touring models became available, and BMW finally stopped making them in June 1996.

This front view of a UK-market E34 shows the broader double-kidney grille which was introduced across the range at the Geneva Show in March 1994.

6 The E34 Super-Saloons and Tourings

The E34s were introduced in a period of economic boom in the West, but before they had been in production for very long that boom turned to recession. All of a sudden, there were fewer customers to be found for high-priced exotic conversions, and the tuning companies found it hard going to sell their re-engineered E34s. Meanwhile, BMW scored heavily with its M5 derivatives, the first one lasting until 1993 and overlapping with the new and more powerful model which entered production in 1991. Those with money to spend took the safe option and bought the Motorsport car, and so this period saw the M5 establish itself even more firmly as the king of the high-performance 5 Series, while there were fewer alternatives to choose from. Nevertheless, the M5 story was not one of unqualified success, and sales tailed off quite markedly towards the end of production.

None of that prevented the introduction of some mouth-watering high-performance E34 conversions, however, and at least one of them proved a strong seller and actually outclassed the M5 ....

THE 3.5-LITRE M5, 1988–93

> We like the M5 because of its character and racing heritage. It's so fast that it makes your eyes water.
> *Motor Trend*, May 1991

The E34 range had already been on sale for two months before its M5 flagship was announced in August 1988. Right-hand drive models took more than a year longer to make their appearance, and were announced at the British Motorfair in October 1989. However, the first of only 200 right-hand drive M5s did not actually reach British showrooms until February 1990. This limited availability reflected the bespoke nature of the car, that would always be a hand-built vehicle rather than a volume-production model. Customers were invited to write their own specification (within reason), and it is probable that no two M5s were exactly the same.

The heart of the new M5 – the second car to bear the name – was a new version of

---

What might have been ...

BMW Motorsport developed an M5 convertible in 1989, a full four-seater with two lengthened doors in the classic cabriolet style. Sadly, BMW cancelled its introduction just a week before it was due to be announced at the Geneva Motor Show. The official reason was that the car's availability would have led to a demand for 5-series convertibles with lesser powertrains, and BMW feared that this would have damaged sales of the popular E36 3-series convertible.

The E34 Super-Saloons and Tourings

Hartge also offered conversions for the E34, the most powerful engine being the 340bhp H5-4.7, based on the 540i. Special exhaust systems, lowered suspension, alloy wheels and aerodynamic aids were all on the menu.

the M88 twin overhead camshaft engine first seen in the M1. A longer stroke increased its swept volume to 3535cc and made it the largest-capacity six-cylinder engine ever made by BMW. And this time, there were no higher-powered versions of the car without catalytic converters. The car had been engineered to run with a three-way catalytic converter as standard equipment, and the size of the Motorsport division's achievement was measurable in a comparison of figures. The 3535cc engine was rated at 315bhp, which was 55bhp more than the previous M5 with catalytic converter, and 10 per cent more than even the non-cat version.

Some careful work had also been done on the intake system, that now had hot-wire air mass metering and a separate throttle for each of the six cylinders. The inlet manifold also featured an ingenious electronic butterfly valve inspired by racing practice that used natural resonance to create a charge effect and so improve throttle response and boost torque by 6 per cent at its peak. Between 3000rpm and 6000rpm, the torque spread was much improved on the new engine, and allowed better use of the performance potential through the five-speed gearbox that was the only transmission available. As on the E28 M5, this gearbox had

A French-market M5 demonstrates the ten-spoke alloy wheels fitted to the last of these magnificent machines.

pre-loaded gears which gave smoother gear selection and reduced noise.

To ensure that the M5 chassis matched the performance potential of this new engine, the Motorsport engineers had changed the brakes, suspension and steering from those on the standard E34 saloons. The ventilated front discs were thicker as well as larger in diameter than on any other 5 Series. The car rode 20mm (0.8in) lower than a 535i, and its suspension featured higher spring rates, different dampers, and fatter anti-roll bars. A self-levelling rear axle ensured a constant ride height and so prevented a loaded boot or rear seat from affecting the handling, and the rear end featured what BMW called 'elasto-kinematics'. This meant that its bushes were compliant enough to allow the whole axle to toe-in during cornering and so generate safe understeer. As for the steering, it was a ZF Servotronic variable-ratio system with a lower ratio than standard to give the driver a greater degree of feel.

By now, BMW knew well enough that the M5's Q-car character was one of its greatest strengths. So the E34 version was visually understated. It came with a front spoiler extension, rear apron and side skirts as standard, but these were all blacked-out and could hardly be seen from a distance. The first cars had no rear spoiler, but one was made optional in November 1988, and was always discreetly painted to match the bodywork. From behind, the polished twin exhaust tailpipes and a black plastic panel between the light clusters distinguished an M5 from lesser E34s, while the front spoiler carried an extra pair of slots that ducted air to the radiator and oil cooler. At extra cost, BMW would black out all the car's chrome to give a look that could be both understated and menacing at the same time.

The aerodynamic aids helped the M5 to retain the 0.32 Cd of the 535i model, despite its fatter tyres – 235/45 ZR 17s made by Pirelli or Michelin. These were carried on unique 8J x 17 M Technic alloy wheels with an appearance so discreet that they were actually too understated for some buyers' tastes. Their two-part design had been developed and patented by BMW. The visible part was a pressure-cast magnesium cover of turbine design, developed in the wind tunnel. Its radial blades gave an axial blower effect to force cold air onto the brake discs and thus maintain maximum brake efficiency at high speeds. The covers were of course handed, to suit the right or left side of the car. Hidden behind the cover was the load-bearing wheel. This was a five-spoke alloy type, and its rim had an asymmetric hump that was designed to stop a deflated tyre rolling off the wheel. M5 customers could specify wider rear wheels with fatter tyres at extra cost.

Inside, buyers could choose from special Motorsport check cloth sports seats with leather side panels or full leather upholstery. Power-adjusted front seats were an option, and could be had with a position memory for a further additional premium. At the rear were individual sports seats, separated by a central armrest with a forward-extending drawer. For extra cost, these could be adjustable, too. They came with rear head restraints as standard, and for extra cost these could be linked to the colour-coded seat belts so that they lowered out of sight when the belts were not fastened (and there was therefore no-one in the rear seats to need them). The headlining on all cars was leather, and

---

M5 with 3.5-litre Engine (1988–93)

Engine: 3535cc six-cylinder (93.4mm bore x 86mm stroke) with twin overhead camshafts and four valves per cylinder, 10.0:1 compression ratio and Bosch Motronic fuel injection; exhaust with catalytic converter and Lambda sensor; 315bhp at 6900rpm and 266lb/ft at 4750rpm.
US models: 310bhp at 6900rpm and 265lb/ft at 4750rpm.
Transmission: Five-speed overdrive manual gearbox (ratios 3.51:1, 2.08:1, 1.35:1, 1.00:1, 0.81:1). Axle ratio 3.91:1; 25 per cent limited-slip differential.
Suspension, steering and brakes: Independent front suspension, with MacPherson struts and anti-roll bar; semi-trailing arm rear suspension with auxiliary pivot link, anti-roll bar and coil-sprung struts attached to the hub carriers.

ZF Servotronic speed-variable recirculating-ball steering with servo assistance.

Servo-assisted brakes, with ventilated discs at the front and solid discs at the rear; dual hydraulic circuit and ABS.

235/45 ZR 17 tyres on 8J x 17 alloy wheels, or optional 255/40 ZR 17 tyres on 9J x 17 alloy wheels.
Dimensions: Overall length 4,720mm (185.8in); overall width 1,751mm (68.9in); overall height 1,392mm (54.8in); wheelbase 2,761mm (108.7in); front track 1,474mm (58in); rear track 1,496mm (58.9in).
Unladen weight: 1720kg (3792lb).

The E34 Super-Saloons and Tourings

Janspeed's twin turbocharged 535i was displayed at the 1989 Motorfair in Britain.

buyers who wanted it could specify a leather lining for the boot instead of the standard woollen one.

The dashboard and controls were special, of course. There was a unique three-spoke steering wheel with moulded thumb grips and the Motorsport logo, that was repeated on the instrument panel between the 300kph (186mph) speedometer and the rev counter with their red needles. The econometer in the bottom segment of the rev counter on other E34s was replaced by an oil temperature gauge – which BMW reasoned was more likely to be of interest to the driver of an M5! A final piece of pure caprice was the illuminated M Technic gear knob. A car telephone and a 12-speaker in-car entertainment system were further options.

All this of course was on top of a comprehensive basic specification that was only to be expected in BMW's flagship. The M5 came as standard with ABS, electric windows, Active Check Control, an on-board computer, limited slip differential, an electric sunroof, air conditioning and headlamp washers.

The specification changed a little in September 1990, when a bench seat was fitted as standard and the individual seats with their fixed central armrest were moved to the options list. The woollen boot trim was also replaced by a velour type, and the boot was equipped with a storage box on each side and a luggage net on its floor. This specification was in place by the time the M5 was released in the USA as a 1991 model. It came with leather upholstery as standard, together with a driver's airbag to add to an already impressive list of standard equipment. Some re-tuning of the engine had been necessary to make the car meet US emissions regulations, and maximum power had dropped to 310bhp. Torque, however, was barely affected, and the car could still reach 60mph from rest in 6.4 seconds.

The M5 was available in the USA for only three years, however, and was withdrawn in 1993. One of the final examples was presented to the State Police in South Carolina, home of the Spartanburg factory that BMW would open in 1994.

THE ALPINA B10, 1989

The original B10 based on the E28 535i was not available for very long. Introduced in 1987, it was only around for a year before the E28 models ceased production in favour of the E34s. But Alpina used the same powertrain in the new E34 to create a second-generation B10 5 Series, this time with a catalytic converter in the exhaust to meet the new German regulations that came into force at the beginning of 1989.

The catalytic converter robbed the engine of some power and torque, with the result that it put out 254bhp at 6000rpm and 225lb/ft of torque at 4000rpm, as compared to the 261bhp and 240lb/ft of the E28 B10 engine. Those figures still made the B10 usefully more potent than the 211bhp, 220 lb/ft 535i in standard trim. However, the E34 was a heavier car than its predecessor, and the greater weight worked together with the lower engine output to give performance that was disappointing when compared with the E28 B10. While the top speed remained similar at around 150mph, the 0-60mph time went up by a big increment, to 7.4 seconds from the 6.1 seconds of the earlier car.

Of course, the E34 B10 did have the latest version of the Alpina suspension, that followed the principles applied to the older E28 models. The coil springs were stiffened, the anti-roll bars were thicker,

---

New on the Block

During the late 1980s, the firm of AC Schnitzer became a force to be reckoned with in tuning equipment for BMW cars. The company was formed by the Kohl Automotive Group in Germany, who until 1986 were exclusive distributors for Hartge products. When Hartge decided to go their own way, Willi Kohl determined to develop his own tuning and customising accessories for BMW cars. A deal was done with Herbert Schnitzer, whose racing team won the European Touring Car Championship that year (with a BMW 635CSi driven by Roberto Ravaglia), and the new range of accessories was shown for the first time at the 1987 Frankfurt Motor Show with the brand-name of AC Schnitzer.

AC Schnitzer sold a variety of accessories, ranging from gearknobs to spoilers, but the company also developed some performance equipment for the E34 5 Series. Options included uprated engines and even complete engine transplants, with the usual complementary array of braking and wheel and tyre changes.

and wider, lower-profile tyres were fitted on Alpina's spoked alloy wheels. This time around, there were 235/45 ZR 17s on 8.5J wheels at the front, with 265/40 ZR 17s on 9.5J wheels at the rear. So even though the E34 B10 may not have been as quick as the E28 version which preceded it, it did offer the stunning handling which was expected of an Alpina product. However, the company knew that it would have to do better if it was to retain its crown as king of the BMW tuners, and turned to a further development of the B10.

THE ALPINA B10 BI-TURBO, 1989

> The road continually runs out before the acceleration.
> *BMW Car*, December 1997

Nearly a decade earlier, Alpina's B7S had graphically demonstrated both the advantage and the drawback of turbocharging. The advantage was a massive increase in power; the drawback a loss of refinement caused by the sudden transition from off-boost to boost. As the success of the M5 had shown, buyers of this kind of car wanted refinement, so Alpina had avoided turbochargers and had stuck to conventional tuning methods for most of the rest of the 1980s. However, turbocharger technology had come a long way during that decade, and by the end of the 1980s it looked as if it might present Alpina with a way out of the performance setback they had encountered with the E34 B10.

The B10 Bi-Turbo had the regular Alpina suspension package and alloy wheels for the E34 models, plus Alpina's own level control system. It had a deep front spoiler and a rear wing on the boot

An unmistakable badge, on the boot lid of the near-legendary Alpina B10 Bi-Turbo.

lid, and there was the usual distinctive front and rear badging. In this case, however, front and rear badges were different: the one on the grille simply read 'Bi-Turbo', while the one on the boot lid carried the full 'B10 Bi-Turbo' name. The traditional Alpina decals were available for the car's sides and for the front air dam, but many customers ordered their cars without them.

The biggest drawback of the early turbocharged cars had been a noticeable delay between the driver's demand on the accelerator pedal and the engine's response. Part of this was caused by the need for extra exhaust gases to pass through the turbocharger before that could spin any faster to provide boost, and part of it was caused by the inertia of the turbocharger itself. The bigger the turbocharger, the more inertia had to be overcome, and therefore the longer the delay, or 'turbo lag'. To overcome the second of these problems, Alpina had fitted the B10 Bi-Turbo with two smaller-diameter turbochargers in parallel. These gave the same amount of boost as a single larger-diameter unit, but with very much reduced lag.

The turbochargers were Garrett T25s, both of them on the exhaust side of the engine and therefore hidden underneath its canted cylinder block. These were supplemented by a huge air-to-air intercooler ahead of the radiator, and there was a control on the dashboard that allowed the amount of boost to be varied between 1.4 bar and 1.8 bar.

With the turbochargers giving their maximum boost, the twin turbocharged engine boasted 360bhp at 6000rpm and 385lb/ft at 4000rpm – figures which completely eclipsed any other E34 conversion before or since. Alpina had feared for the durability of the standard Getrag 280 five-speed gearbox with all that torque being put through it on a regular basis, and so in its place the B10 Bi-Turbo had a stronger Getrag 290 five-speed, which brought a particularly sweet change as a side-benefit.

Maximum performance was simply shattering. The B10 Bi-Turbo would storm to 62mph (100kph) from rest in just 5.2 seconds, and carry on right up to a maximum speed of 181mph (291kph). It is

The reverse-angled spokes of the 3.8-litre M5's wheels contributed to brake cooling.

perhaps unfair to compare that maximum with the much lower maximum of an M5, because the factory-built car was fitted with a limiter that prevented it from exceeding 250kph (155mph) as part of a gentleman's agreement among the leading German car makers. However, the other figures tell their own story.

The Bi-Turbo could of course be kitted out with all kinds of interior extras from the Alpina list, and was just as much of a bespoke car as the M5 itself. Despite its enormous cost, even in Germany, it attracted a good number of customers, too. In the five years of its availability, between 1989 and 1994, no fewer than 507 examples were built. All of them are already coveted by lovers of fine sporting machinery.

THE JANSPEED TWIN-TURBOCHARGED 535i, 1989

Alpina were not alone in recognising that turbocharging had come a long way since the beginning of the 1980s. In Britain, turbocharging pioneers Janspeed Engineering of Salisbury had also decided to see what could be done with twin turbochargers on an E34, and their conversion of a 535i was announced at the London Motorfair in October 1989. The show car was equipped with a bodykit of side sills and front and rear spoilers, the horizontal grilles beside its headlights were finished in the body colour, and it sported chrome-finish wheels with low profile tyres.

Janspeed claimed 'a massive increase in both power and torque' for this car, but unfortunately it has not been possible to discover exactly what the figures were. Nor is it clear at present how many examples of the car were made.

THE 3.8-LITRE M5, 1991-1995

Few people doubted that the M5 was the world's best sports saloon, but BMW was not about to rest on its laurels. There was new pressure from long-standing rivals Mercedes-Benz, whose new 326bhp 500E was offering 0–60mph times of 6.3 seconds with an automatic transmission, and

---

M5 with 3.8-litre engine; saloon (1991–95) and Touring (1992–95)

As for M5 with 3.5-litre engine, except:
Engine: 3795cc six-cylinder (94.6mm bore x 90mm stroke) with twin overhead camshafts and four valves per cylinder, 10.5:1 compression ratio and Bosch Motronic M3.3 engine management system; exhaust with catalytic converter and Lambda sensor; 340bhp at 6900rpm and 295lb/ft at 4750rpm.
Austrian and Swiss models: 334bhp; later 327bhp.
Transmission: Five-speed overdrive manual gearbox to April 1994 with ratios as for 3.5-litre car. Six-speed overdrive manual gearbox from May 1994 (ratios 4.23:1, 2.52:1, 1.66:1, 1.22:1, 1.00:1, 0.83:1). Final drive 3.23:1 from May 1994.
Wheels and tyres: 9J x17 alloy wheels with 255/40 ZR 17 tyres optional at the rear on saloons up to April 1994 and standard on Touring models before that date. From May 1994: 245/40 ZR 18 tyres on 8J x 18 alloy wheels at the front and 9J x 18 alloy wheels at the rear.
Unladen weight: 1800kg (3968lb) for Touring.

Both handling and ride quality improved on the 3.8-litre M5, thanks to computer-controlled Adaptive M-Technic suspension. This model is surely one of the most desirable 5 Series ever made.

of course the Alpina B10 Bi-Turbo was also providing crushingly superior performance. So BMW Motorsport started work on an even more powerful M5, which was previewed at the Frankfurt Motor Show. Taking advantage of an unexploited niche in the market, the Bavarians decided to offer the M5 package in the Touring bodyshell as well as in the saloon, and this was also previewed at the 1991 Frankfurt Show although it did not go on sale until spring 1992 and even then was not offered in all markets.

The new M5s were well-received, but they did not sell as well as the older cars with which their production overlapped for a couple of years. One reason must have been that they were never available in the USA. In cold figures, however, nearly twice as many of the earlier cars were made in six years as were made of the later saloons and Tourings combined in five years.

So the M5 remained a low-volume model, still hand-built at Garching. In Britain, for example, only 60 were imported during its first year of production.

The centrepiece of the revised car was a new version of the four-valve big six engine, bored and stroked to give a swept volume of 3.8 litres. Larger valves and inlet ports, a higher compression ratio and the newly-developed Motronic M3.3 engine management system helped to boost power

The E34 Super-Saloons and Tourings

Left: The discreet boot-lid badge incorporated a flash with the Motorsport colours.

Above: With 340bhp, the 3.8-litre engine offered stunning performance. Note the different cam covers from the 3.5-litre type shown in the colour section.

by 8 per cent up to 340bhp and torque by 11 per cent up to 294 lb/ft. Perhaps more important was that there was also much more torque low down the rev range, with 75 per cent (221 lb/ft) available at just 1800rpm. This not only made the engine more flexible in performance terms, but also made for smoother driving in stop-go traffic. That smoothness had received further attention in the shape of a lightened clutch, while the new Motronic 3.3 engine management system (seen earlier in the 750i limousine) helped damp out the jerkiness of stop–go driving. The 3.8-litre engine also featured a refined version of the electronic secondary throttle butterfly valve system that was now integrated into the engine management system, and for surer ignition there was one solid-state high-tension coil for each of its six cylinders.

The result was to make the world's best sports saloon even more docile in city traffic while improving its acceleration and its maximum and cruising speeds. BMW's own claims were of 0–62mph (100kph) acceleration in 5.9 seconds and a 6.9 second 50–75mph time in fourth gear. Top speed remained electronically limited to 155mph (250kph), although 170mph (273 kph) would otherwise have been within easy reach. The Motorsport division had developed a pair of special all-metal exhaust catalysts for the car and, as before, these were fitted to M5s for all markets. However, there were lower-powered versions for Switzerland and Austria, whose exhaust emissions regulations were too tight for the standard car to meet.

For this third-generation M5, BMW Motorsport had also developed Adaptive

M-Technic suspension, which was a special Motorsport version of the third-generation Electronic Damper Control (EDC III). This was a fully automatic system that took account of road speed, steering input, acceleration, retardation and vertical body movement front and rear to programme within a split second the most suitable of three settings for the dampers. This meant that relatively supple damping for bumpy roads could be transformed almost instantly into stiffer shock absorption when, for example, turning into a corner quickly or swerving to avoid an accident.

For owners who wanted to enhance the handling of their M5s even further, there was the 'Nürburgring' package, named after Germany's famous race circuit. This brought wider 9J x 17 rear wheels fitted with 255/40 ZR 17 tyres, a thicker rear anti-roll bar, sports-tuned Servotronic steering and a switch for the adaptive suspension to maintain firmer settings throughout the entire driving range.

From the outside, the 3.8-litre M5 remained as understated as its predecessor. Only new chunky five-spoke forged alloy wheels and two unique M5 colours (Avus Blue and Daytona Violet) revealed the car's identity from a distance, but this time there was also a band of contrasting colour all round the bottom of the car, taking in the sills, the lower front spoiler and the aerodynamic lip below the rear bumper. A closer look would of course also reveal the familiar M5 badges on the grille and boot lid. Inside, the seats were upholstered in a new Motorsport striped cloth in grey or black. The headrest coverings and seat edgings were in easy-care Amaretta Suede, but a full leather interior remained an option.

The M5 followed developments in the rest of the E34 range to some extent, taking on new mirrors, side impact bars, seat belt pre-tensioners and the new engine immobilizer in September 1992. From September 1993, a driver's airbag was also standardized for all markets. Then the early spring of 1994 brought a facelift with the broader grille introduced on other E34s at the same time, and the M5 was also fitted with strikingly attractive new ten-spoke alloy wheels of 18-inch diameter. New 'floating' brake discs were fitted at the same time – a type of disc usually seen only on racing cars – with multi-part construction that offered better braking and longer life thanks to better heat dissipation and lower weight. The fatter rear anti-roll bar of the Nürburgring suspension was now standardized, together with the latest six-speed manual gearbox allied to a taller final drive. The revised cars benefited from increased flexibility, with 50–70mph

---

Hartge conversions of the E34

Hartge once again offered a range of performance options for the 5 Series range. These could be accompanied by the customer's choice of exhaust systems, suspension kits, uprated brakes, alloy wheels (in 16, 17, 18 and 19-inch sizes), and body addenda. There was also a range of limited-slip differentials, and alternative axle ratios for the pre-1990 535i (3.73:1 for the manual models and 3.91:1 for the automatics).

The performance tuning started with the 231PS H5-2.8, based on the four-valve 525i. For the four-wheel drive 525iX, the H5-3.0 conversion brought 250PS. A head conversion on the 535i was called the H5-3.4 and brought power up to 245PS, while the H5-3.6 for the 530i offered 270PS and the H5-4.7 for the 540i took power up to 340PS.

The E34 Super-Saloons and Tourings

The Alpina B10 Bi-Turbo was a giant among super-saloons. This one was pictured by Dave Shepherd.

acceleration of just 6.7 seconds in fourth gear, and fuel consumption averaging a very reasonable 25mpg.

Most M5s were more or less built to order, but right at the end of the production run BMW made a small number of very special cars. In Britain, the finish of right-hand drive M5 production was commemorated with a limited edition of 15 cars finished in British Racing Green with leather upholstery. The very last car was made in July 1995 – and there would be a gap of three years before its successor was announced.

THE MST-WIESMANN CONVERSIONS, 1993 ON

MST (Motor Sport Technik) was established in 1993 at Dulmen in Germany by Karl-Otto Noelle, who had until then been head of development at Alpina. In order to market his conversions, he teamed up with Wiesmann, makers of a special BMW-based roadster, and the company became known as MST-Wiesmann.

Noelle started in a small way with engine and suspension upgrades for the six-cylinder E34s. Broadly similar conversions were available for the later E39 models as well, and fuller details of these conversions will be found in Chapter 8.

THE ALPINA B10 4.6, 1994

BMW had already signalled its intention of phasing out the big-block six-cylinder engines when the 530i was withdrawn in 1990, and just a year later the 3.5-litre engine on which Alpina's conversions were based went out of production as well. Supplies of engines would be guaranteed for a time, but it was clear that Alpina had to move on.

What was by now quite clear was that BMW planned to use vee configurations for its larger-capacity engines in the foreseeable future. There had been a V12 in production since 1987, initially for the E32 7 Series and later for the E31 8 Series coupés, and from 1992 there were the two new V8s that went into the 5 Series, 7 Series and 8 Series cars. Alpina had its own 350bhp B12 version of the 5-litre V12 engine ready in 1989, and made that available in the big 7 Series saloons and subsequently in the 8 Series coupés. As soon as the V8s became available, the Buchloe company started looking at those, too, and by 1994 it was ready to announce a new high-performance E34 derivative.

The new car was called the B10 4.6, the B10 designation being retained to pick up on the success of the legendary Bi-Turbo and the 4.6 indicating the engine's capacity in litres. The swept volume of the latest 4-litre V8 had been increased to give 4619cc (bore was 93mm and stroke 85mm), and the maximum power was up from the standard engine's 286bhp to a huge 340bhp at 5700rpm. Torque climbed from 289lb/ft to 353lb/ft at 3900rpm – usefully lower down the rev band than on the standard engine. The B10 4.6 of course had the usual Alpina suspension modifications and spoked alloy wheels, and it carried a deeper front air dam and a hoop spoiler on the tail.

Performance was excellent, although the B10 4.6 was faced with formidable opposition from the contemporary 3.8-litre factory-built M5. With a claimed 0–62mph (0–100kph) time of 6.3 seconds, it had no performance advantage off the line when compared to the Motorsport car. Where it did score, however, was in a higher top speed (claimed to be 'over 167mph' (269kph+)) and in the fact that its V8 engine had a very different character from the highly-tuned four-valve six in the M5.

Performance Comparison

N.B. These figures are for European-specification cars and are intended as a guide only.

Model	Power bhp	Torque lb/ft	0-60mph	Max. speed
BMW M5	315bhp	266 lb/ft	6.4 sec	155mph/249kph
Alpina B10	254bhp	225 lb/ft	7.4 sec	132mph/212kph
B10 Bi-Turbo	360bhp	385 lb/ft	5.2 sec (*)	181mph/291kph
M5 (3.8)	340bhp	295 lb/ft	5.9 sec (*)	155mph/249kph
B10 4.6	340bhp	353 lb/ft	6.3 sec (*)	167mph+/269kph+

(*) Figures for 0–100kph (0–62mph).

7 Fourth Generation: The E39 Models From 1995

The development of the E39 5 Series models represented an enormous investment for BMW, a claimed 1,000 million Deutschmarks or £446 million at late-1995 exchange rates. The main reason for this huge cost was of course that the 5 Series was already widely acknowledged as the benchmark medium-sized saloon, and so the standards of the new one had to be even higher if BMW was not going to lose its reputation. And, of course, the 5 Series was central to BMW's whole product programme. So, beginning in the early 1990s, Munich's engineers were given a brief to make the new car ride better, handle better, go faster, use less fuel and be safer in an accident than the E34. It represented a formidable challenge.

The fact that the target buyers for the 5 Series were notoriously conservative in their tastes made this brief even tougher. In making the transition from the E12 to the E28 models ten years earlier, BMW had erred too much on the side of caution, and many people had felt that the incoming models were not different enough from the

The E39 was an imposing-looking car from most angles. The headlamps behind glass and the grille integrated into the bonnet panel were new to the 5-series, but the car was still unmistakably a BMW. This is a top-model 540i, with the 4.4-litre V8 engine.

Fourth Generation: The E39 Models from 1995

Left: Interior styling was beyond reproach, as this UK-market E39 shows.

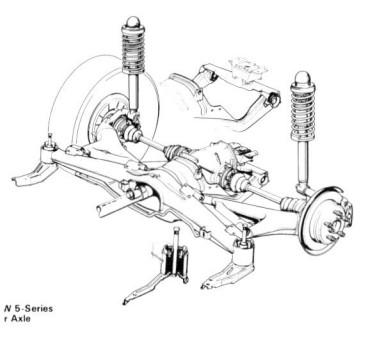

Above: A multi-link rear suspension promised exemplary ride and handling.

outgoing models. However, when the E34s had replaced the E28s, the company had got the balance right. So the approach this time around had to be much as before, and the new cars were in many ways evolutionary developments of the old. Yet there was so much new technology in them that they were undeniably revolutionary at the same time.

With the E39s, launched in October 1995, BMW got it right once again. The cars arrived hot on the heels of the new E-class medium-sized saloons from Mercedes-Benz, which as always proved an extremely hard act to follow. Once the new E-class had been launched, BMW cunningly issued some 'taster' pictures of the new 5 Series to the press, as a reminder that their challenger would be available in a few months' time, and so the motoring world knew what the E39 would look like as early as May 1995. And neither BMW nor the motoring public was disappointed when the press got their hands on test cars.

The E39s were once again acknowledged as the best in their class – superior, even, to the E-class Mercedes – and as raising the standards in the medium-saloon sector. Over the next few months, the E39s also raked in award after award from the motor industry and motoring press for their excellence.

The E39's new shape had been developed under BMW's styling chief Chris Bangle. As was to be expected, it picked up

---

Production of the E39 5-Series

The E39 models were built at the same Dingolfing plant which had produced their E34 predecessors. An integrated production system allowed for saloon and Touring models to be built on the same assembly lines. At 1997 production rates, the Dingolfing plant was able to turn out up to 1,000 cars a day, of which 200 (20per cent) were Touring derivatives.

143

Fourth Generation: The E39 Models From 1995

The rear three-quarters view was perhaps the most unsatisfactory, and the E39's styling could look rather heavy from this angle. Despite the short rear deck, the boot was capacious, and the low sill made it easy to load as well.

design cues from the established E36 3 Series cars and from the recently announced E38 7 Series limousines. These included the sleek nose with headlamps and turn indicators behind flush-fitting glass lenses. The cars had an imposing appearance, though they were somehow bulkier and less svelte than the E34. From the more rounded nose presented by the latest version of the traditional twin-kidney grille, the lines of the bonnet swept back in a broad and purposeful 'power bulge' to join the steeply-raked windscreen pillars. That grille was now integrated with the bonnet pressing and lifted up with it, as the grille had for so long on rival Mercedes products. The passenger compartment had the traditional BMW dog's-leg styling feature at the trailing edge of the rear door window, and a curvaceous rear window pillar led down to the boot. Careful use of styling features on the body sides disguised the fact thAt the car had a wedge-shaped profile with its boot higher than its bonnet, and there were neat touches in the oval door handles and in the tail-lamp clusters that hinted at the styling of the 3 Series.

It came as no surprise to discover that the E39 was bigger than its predecessor. It was longer by 55mm (2.2in), wider by 49 mm (1.9in), and taller by 23mm (0.9in), and there was a longer wheelbase incorporated in those dimensions. The extra 69mm (2.7in) between axle centres brought a welcome increase of an inch in legroom for the rear passengers, while the extra width made for extra shoulder-room all round. Nevertheless, the slippery shape made the car 20 per cent more aerodynamic than its predecessor, to the benefit of fuel consumption at speed. The bodyshell as a whole was also a massive 82 per cent more rigid than before, thanks to its use of high-

Fourth Generation: The E39 Models From 1995

strength steel and aluminium. This extra rigidity improved the whole feel of the car on the road, but it also made a major contribution to its crashworthiness. As before, the front and rear sections were designed to deform progressively under impact around a rigid passenger cell, and in fact that cell would remain intact during a 35mph offset frontal collision, thus exceeding all safety legislation then current. For good measure, and to meet US regulations, the doors incorporated diagonal aluminium bars to guard against side impacts. Yet, incredibly, the bodyshell was no heavier than that of the E34s.

A great deal of weight – 65kg (143lb) per car – was saved by using aluminium alloy for the suspension components. So steel was absent from all the tie rods, track control arms, hubs, outer strut tubes, brake callipers and sub-frames, and even the road wheels were made of aluminium alloy. This brought benefits not only to the overall weight of the car but also to the ride and handling through a lower unsprung weight. Careful positioning of components and detail measures such as reducing the weight of the final-drive assembly resulted in a near-ideal 50/50 weight distribution between front and rear, that made a further important contribution to ride comfort and to handling. There was still noticeable roll in fast cornering, however, and BMW had developed a sports suspension option – developed by M-Technic – which lowered the car by 25mm and reduced this to more sporting levels.

While this all-aluminium suspension was the first of its kind on a volume production car, the steering was a first for a 5 Series BMW. It was a rack-and-pinion system mounted ahead of the front axle line with a double-jointed axle incorporating tie-rods. Earlier 5 Series had stuck resolutely to recirculating-ball

Aerodynamics was one reason why the headlamps and indicators of the E39 were shielded behind these elegant covers.

Note the boot lock, offset to one side. The tail lamp clusters did not carry over the characteristic 'step' of the E34 styling.

A high-mounted third brake light was standard on all E39 models, and was neatly integrated into the rear window and parcels shelf.

Fourth Generation: The E39 Models From 1995

The boot was sensibly shaped, and the spare wheel lived below its floor. Elasticated luggage retaining straps were a useful feature.

Underbonnet presentation was immaculate, as always. This is the V8 engine in a 535i.

On opening the door, the E39 driver was greeted by this neat kick-plate with the BMW name.

Luxury in a door trim! Note the two large stowage pockets, the padded leather insert and the wooden fillet.

The neat instrument binnacle and switchgear were another BMW triumph.

The leather seats were neatly and attractively finished, and the leather gaiter around the handbrake lever added a touch of class.

Fourth Generation: The E39 Models From 1995

This is the Switchtronic transmission control on a right-hand drive automatic 535i. On the right of the gate, P, R and N give the normal functions.

steering, even for the high-performance M5 variants. The positive feel that the new system brought to the steering was aided by a version of the semi-active elasto-kinematic rear suspension already seen on the E34 M5 and on current 7 Series limousines and 8 Series coupés. This had been designed by the engineers in BMW's M-Technik 'think-tank', and its effect was to promote safe understeer. The rear suspension also incorporated pneumatic self-levelling struts and the rear wheels had ASC + T (Automatic Stability Control and Traction) as standard – although this could be switched off if the driver preferred. In addition, EDC-III damper control, already seen on the M5, 7 Series and 8 Series, was optional; this was not available on the very first cars, however, because its introduction was delayed until early in 1996. As for the brakes, the E39s had the latest-generation ABS 5 system, with discs that were claimed to be fade-free. Stopping power was indeed extraordinarily good.

The driver was faced with the usual clear set of dials and with instruments and controls that were delightful in their precision. An airbag was incorporated as standard in the steering wheel hub, and a passenger airbag was also standard right across the range. The dashboard was the usual neat and stylish moulding, and actually broke new ground with its all-polyurethane construction; traditionally, dashboards were made up of several different plastics, but BMW's aim had been to make this one easily recyclable.

Within the instrument panel, a window displayed pictograms to warn of open doors, failed exterior lamp bulbs, low washer-bottle fluid, and the like. When an on-board computer was fitted, it was incorporated in the centre console below the controls for the sophisticated new thermostatically controlled heating and ventilating system that included a pollen filter in the air intake as standard. An especially ingenious feature of this was the innovative latent-heat reservoir in the front passenger footwell. This stored waste heat from the engine, that would be supplied to the heater within a few

---

BMW in Cairo

On 18 June 1997, BMW opened a new CKD assembly plant near Cairo in Egypt. It was the company's first CKD plant in the Middle East. The car chosen for production start-up was the E39 523i, which was shipped out in component form from Germany. The Egyptian-built cars nevertheless had 40 per cent of their components locally sourced.

seconds of start-up so that heat was available long before the engine was fully warm. It also allowed the heater to be run for a time while the engine was turned off, and it would feed heat back into the engine itself to promote faster warm-up after a cold start.

Air conditioning could of course be added as well. Also in the console was the RDS radio that drew signals from an antenna incorporated in the rear window demister elements. The radio and its accompanying cassette deck, concealed under a hinged flap, were non-standard sizes unique to the 5 Series in order to discourage thieves. They were one more enhancement in the BMW security systems, that already included as standard an alarm, deadlocks and the EWS II transponder ignition key system.

Upholstery came in three types of fabric and three types of leather. The fabrics were striped flock, wool velour or a high-quality flat-weave rattan cloth. The leather options were Montana, Nappa, or buffalo hide. Front seats on all models came with electric adjustment, and seat heaters were optional. The top seat option was known as the Comfort seat, and these seats had adjustable shoulder sections. They were of course standard on the more expensive models. As for the rear, the seat could be had with a one-third/two-thirds split folding backrest to give more load-carrying space, or it could be bought with fold-out child seats. These could also be retro-fitted to the car on request.

BMW had come up with an incredible array of options and features to enable each customer to tailor a car to his or her individual taste. Among them were items like the AIC rain-sensor wipers (that adjusted the wiper speed automatically to suit the level of rainfall), the fold-out zipped bag which would take two pairs of adult skis, and the multifunction steering wheel. This astonishing device was already available on the 7 Series limousine and allowed the driver to control the radio, cruise control, heating and ventilating, and built-in mobile phone from buttons in the wheel. For cold climates, it was possible to order a heated steering-wheel rim, too. The list seemed endless, and of course every

Performance Comparison

All figures provided by BMW.

Model	0–62mph (0–100kph)	Maximum speed
520i manual	10.2 sec	137mph/220kph
520i automatic	11.2 sec	130mph/209kph
523i manual	8.5 sec	142mph/228kph
523i automatic	9.6 sec	140mph/225kph
525tds manual	10.4 sec	131mph/211kph
525tds automatic	11.0 sec	127mph/204kph
528i manual	7.5 sec	147mph/236kph
528i automatic	8.8 sec	145mph/233kph
535i manual	7.0 sec	153mph/246kph
535i automatic	7.7 sec	152mph/245kph
540i manual	6.2 sec	155mph (limited)/249kph
540i automatic	6.5 sec	155mph (limited)/249kph

item on it served to inflate the cost of the basic car. However, the popular myth that BMWs came with spartan levels of equipment and that every feature worth having cost extra was simply not true. Even in standard form, the E39s were comprehensively kitted out.

THE 523i, 525tds AND 528i, 1995

> The 525tds is quite simply the best diesel car on the market. No question about it.
> *BMW Car*, February 1997

BMW did not launch the entire E39 range at once in the autumn of 1995. In order to allow a more gradual changeover from the outgoing E34s, it introduced only the two mid-range petrol models and the turbodiesel at first. This choice of models was probably made to protect sales of slower-selling entry-level E34s (518i and 520i) and of the top-of-the-range cars as well: dealer stocks of the best-selling mid-range models would have been low and demand would have been high and instantaneous for the new cars.

The new cars were introduced at the Frankfurt Show in September 1995, but the first examples did not become available through dealers until December. Right-hand drive cars followed a few months later, as did the remaining three models of the launch range and a number of additional options. Meanwhile, the E34 Touring models remained on sale, and their production did not end until June 1996. The E39 Touring models followed shortly afterwards.

BMW model designations had indicated the engine size for more than 30 years, and the only exceptions in the 5 Series range had been the E28 525e and US-model 528e with their 2.7-litre engines. With the E39s,

Top: The new 2.5-litre turbocharged diesel engine was available in two outputs: 115bhp for the 3 series and (below) with intercooler and 143 bhp for the 5 series.

market pressures forced another exception to the rule. The new 2.5-litre BMW, successor to the 525i, was now badged as a 523i. The main reason was that the car was intended to compete with the 2.3-litre Mercedes-Benz E230 and BMW were anxious to make clear that it was not significantly more expenive simply because of its engine size. However, it was also true that the new designation avoided confusion between 525i petrol model and 525tds turbodiesel.

So the 523i had the latest version of the

Fourth Generation: The E39 Models From 1995

This side view of the Touring – in fact, a 540i model – shows the model's slightly longer rear overhang as compared to the saloons. Note how the rear door window frames had been redesigned to suit the Touring's roofline.

There was a suspicion of heaviness about the Touring from some angles, and this front view of a 540i certainly makes the car look less svelte than its E34 predecessor.

2494cc M50 four-valve six-cylinder, boasting 170bhp instead of the 192bhp it had offered in the E34 525i. BMW's reasoning was that there was no need for the extra power if the car was competing with Mercedes' smaller-engined model, and the advantage of the lower power output was that it contributed to fuel savings of around 6 per cent. Meanwhile, the engine in the 528i was a new 2793cc six-cylinder that had made its debut in the E36 328i. Once again, it was a member of the M50 family, with twin overhead camshafts and four valves per cylinder. In this case, it offered 193bhp – just a little more than the outgoing 525i. One important reason for pegging the power at this level was that it met a break point in German insurance categories.

These power outputs sounded disappointing to someone who had not understood what BMW was trying to do. Outright power was not the issue when the 523i could reach 142mph and the 528i peaked at 147mph. What the BMW engineers had done was to concentrate their efforts on low-speed torque to give their engines effortless acceleration and better fuel economy. While the 2.8-litre engine can perhaps not be directly compared with any predecessor, a comparison of the torque figures for the 2.5-litre engine in the E39 and E34 guise is revealing. The E34 525i engine had 180lb/ft at 4500rpm, while the E39 version in the 523i put out very slightly more torque (181lb/ft) at a significantly lower engine speed (3950rpm).

As for the 525tds, its engine was broadly unchanged from the 143bhp one that had performed so successfully in the E34 model of the same name. All three models came as standard with a five-speed close-ratio gearbox with direct top gear, and the option was an overdrive five-speed automatic with AGS.

This Adaptive Gearbox System automatically adjusted the change-up points to the driver's style, choosing

---

Engines in the E39 Models

The engines in the E39 models were all six-cylinder or V8 types. For the first time, there was no four-cylinder model in the range.

The four-valve sixes
There were three varieties of the M52 four-valve six-cylinder in the E39 range. These were in the 520i (1991cc), 523i (2494cc) and 528i (2793cc). All of them had twin overhead camshafts driven by roller chains.

The turbodiesel
The M51 turbodiesel in the 525tds models had a capacity of 2498cc. It had two valves per cylinder and a single overhead camshaft, with indirect injection, an air-to-air intercooler and Bosch DDE II engine management.

The V8s
The M62 V8 engines were further-developed versions of the smaller-capacity M60 engines seen in the E34 5 Series. They came with a 3498cc capacity for the 535i and a 4398cc capacity for the 540i. Both were all-alloy engines with four chain-driven overhead camshafts, hydraulic tappets and four valves per cylinder.

Fourth Generation: The E39 Models From 1995

from a range of two Sports programmes for fast driving, two Economy modes for more leisurely work, and a fifth which adapted itself to suit any style not encompassed by the other four.

THE 520i, 535i AND 540i, 1996

> The 535i is not only a slick act that's hard to fault, it also inspires true devotion.
> *BMW Car*, July 1997

This time around, BMW had decided firmly against putting a four-cylinder engine into the entry-level model of the 5 Series. So there was no 518i; instead, the six-cylinder 520i became the entry-level car. It was announced in May 1996 together with the two new top models of the E39 range. These were both V8-powered cars, and carried the designations of 535i and 540i.

The 520i brought no real surprises. Its engine was the four-valve 2-litre six from the outgoing E34 model, still with 150bhp but with more bottom-end torque at lower crankshaft speeds – exactly as in the 2.5-litre 523i. However, the two V8 engines were very different from those seen in the E34s. Now bearing the M62 designation, they were larger-capacity developments of the originals and had been introduced in the E38 7 Series limousines shortly before their appearance in the E39 range. These engines were heavier than the six-cylinder types, and so the front axle carrier on the V8-engined E39s was made of steel rather than aluminium as on the other models. In addition, the two V8 cars had recirculating-ball steering rather than the rack-and-pinion type of the others. It saved space and, according to BMW, it also gave a better steering feel.

The old 3-litre V8 had been developed into a 3.5-litre engine by means of a

The optional extendable load floor was a very welcome addition to the Touring, and is seen here in one of its many practical guises. For buyers with bad backs, it was a godsend!

Fourth Generation: The E39 Models From 1995

Above: The passenger compartment of all the E39s was a cosseting place to be.

The rear seats offered high levels of comfort – and safety – for three passengers.

A multi-function steering wheel was optional, as seen here.

Fourth Generation: The E39 Models From 1995

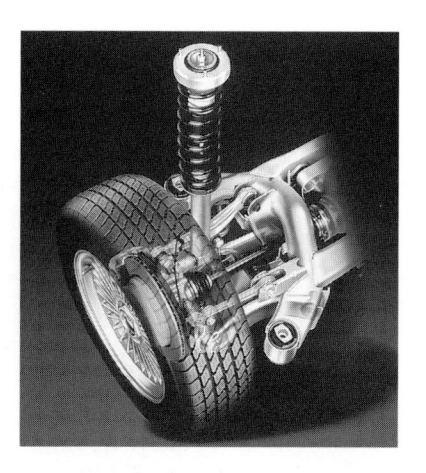

Above: The saloon's rear suspension was of course mounted on its own sub-frame.

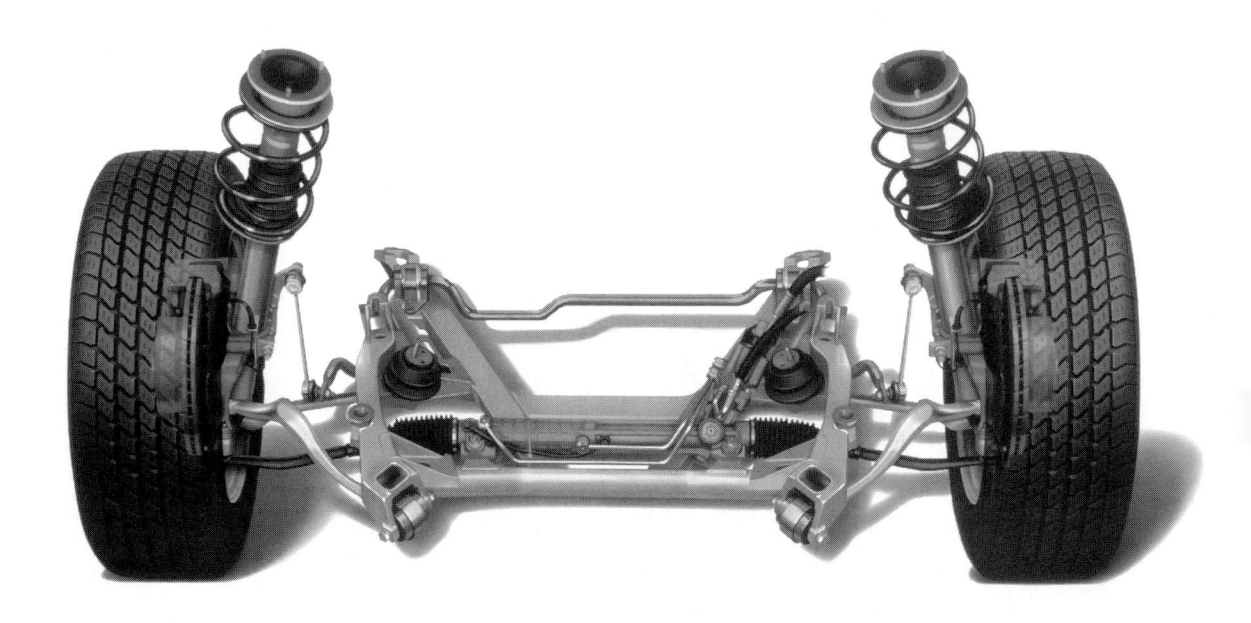

Above: Touring models had a more compact rear suspension to allow for a flat load floor.

The alloy front suspension of the E39 was new, but still featured BMW's favoured spring struts.

Fourth Generation: The E39 Models From 1995

lengthened stroke, while the old 4-litre V8 had been bored and stroked to give a capacity of 4.4 litres. The 3.5-litre-engined car logically took on the 535i name, but BMW confused the issue by retaining the old 540i designation for the 4.4-litre car. Once again, the Munich engineers' aim had been to improve low-speed torque, and in the smaller engine that improvement had been astounding. Torque of the 3-litre engine had been 209lb/ft at 4500rpm, while torque in the new 3.5-litre had increased substanially to 236lb/ft at the much lower crankshaft speed of 3300rpm. This gave the engine a muscular flexibility which had simply not been present in its predecessor. As for the larger V8, peak power remained unchanged at 286bhp (although this was at 5700rpm rather than 5800rpm) while torque went up from the 4-litre's 289bhp at 4500rpm to 310lb/ft at 3900rpm.

There had been other changes, too. The 3.5-litre V8 now had a forged steel crankshaft like its bigger brother in place of the 3-litre's cast-iron item. Valvetrain changes had reduced the reciprocating weight, and single-row roller chains for the camshaft drives had replaced the heavier duplex chains of the older engines. The timing-chain idler gear at the base of the engine's vee had been replaced by a weight-saving aluminium guide rail with a low-friction plastic track clipped to it, which lso helped reduce noise. Cam profiles had been altered to give lower lift (to reduce friction) and less overlap, and work on the intake system and inlet ports had more than compensated for the consequent poorer cylinder filling.

Cooling system changes had also helped improve fuel economy by allowing the engine to run at optimum termperature for more of the time. What BMW described as a demand-oriented cooling system used an electronically heated thermostat linked to

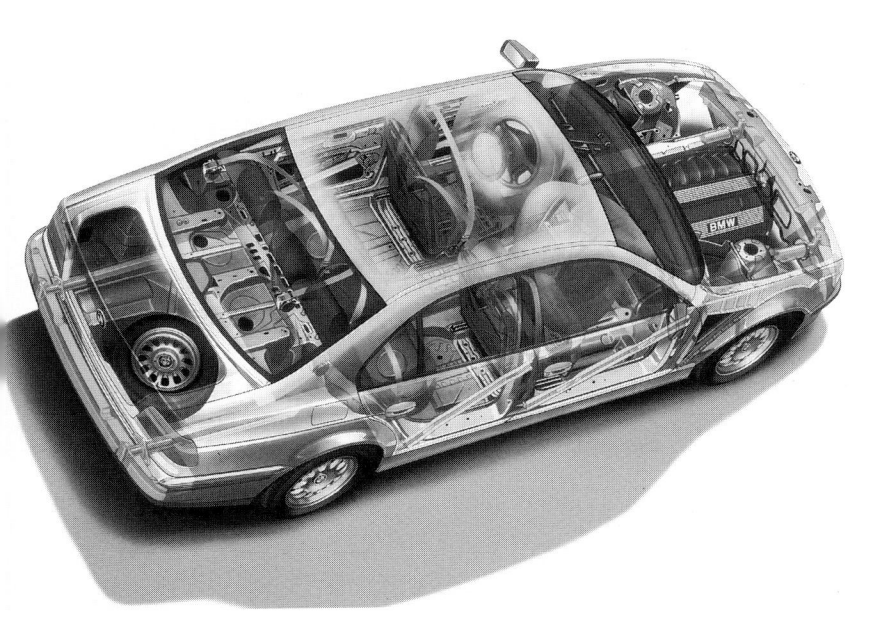

Above: BMW 5 series passive safety measures are as good as any in the motor industry. Twin airbags are fitted as standard equipment.

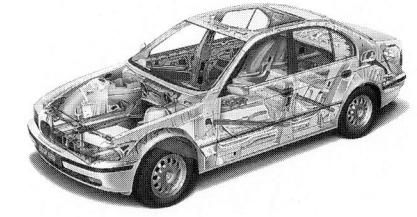

Above: A cutaway drawing showing BMW 5 series high-strength components.

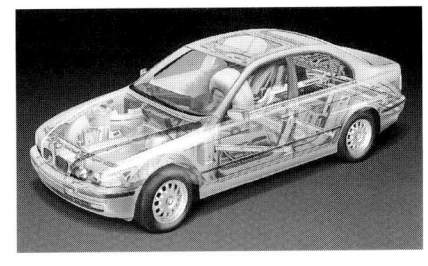

Above: A cutaway showing the BMW 5 Series safety body.

Fourth Generation: The E39 Models From 1995

BMW E39 5 Series Models, from 1995

All models shared the same basic architecture of a unitary four/five-seater bodyshell with front and rear crumple zones, with a front-mounted engine driving the rear wheels.

520i SALOON AND TOURING (1996 ON)

Engine:
Cylinders	1991cc six-cylinder
Bore and stroke	80mm x 66mm with twin overhead camshafts and four valves per cylinder
Compression ratio	11.0:1
Carburettor	Bosch Motronic engine management system exhaust with catalytic converter and Lambda sensor
Max power	150bhp at 5900rpm
Max torque	140lb/ft at 4200rpm.

Transmission:
Gearbox	Five-speed close-ratio manual
Top	5.10:1
4th	2.77:1
3rd	1.72:1
2nd	1.22:1
1st	1.00:1

Optional five-speed overdrive automatic with AGS
- 1.366:1
- 2.00:1
- 1.41:1
- 1.00:1
- 0.74:1

Steptronic optional from 1996
- 3.72:1
- 2.04:1
- 1.34:1
- 1.00:1
- 0.80:1

Axle ratio 3.46:1
(automatic 3.64:1)

Suspension and steering

Independent front suspension, with double-joint spring struts, aluminium control arms and anti-roll bar; multi-link rear suspension with coil-springs and anti-roll bar.

Steering	ZF rack-and-pinion steering with servo assistance.
Brakes	Servo-assisted all-disc brakes with dual hydraulic circuit ABS standard.
Tyres	205/65 R 15
Wheels	6.5J x 15 aluminium wheels

or:	225/60 R 15 tyres
	7J x 15 alloy wheels
	16–inch and 17–inch alloy wheels optional.

Dimensions

Overall length	4,775mm (188in)
Overall width	1,800mm (70.8in)
Overall height	1,435mm (56.5in)
Wheelbase	2,830mm (111.4in)
Front track	1,512-1,516mm (59.5–59.7in), depending on wheel and tyre choice
Rear track	1,526-1,530mm (60–60.2in), depending on wheel and tyre choice.
Unladen weight	1400kg (3086lb).

523i SALOON (1995 ON) AND TOURING (1996 ON)

As for 520i, except:
Engine

Cylinders	2494cc six-cylinder
Bore and stroke	84mm x 75mm with twin overhead camshafts and four valves per cylinder
Compression ratio	10.5:1
Carburettor	Bosch Motronic engine management system exhaust with catalytic converter and Lambda sensor
Max power	170bhp at 5500rpm
Max torque	181lb/ft at 3950rpm.
Unladen weight	1420kg (3130lb).

525tds SALOON (1995 ON) AND TOURING (1996 ON)

As for 520i, except:
Engine:

Cylinders	2498cc six-cylinder
Bore and stroke	80mm x 82.8mm indirect-injection turbocharged diesel with overhead camshaft
Compression ratio	22:1
Carburettor	Garrett T 03 turbocharger, air-to-air intercooler and Bosch DDE injection; oxidation catalyst
Max power	143bhp at 4800rpm
max torque	188lb/ft at 2200rpm.
Transmission	No Steptronic option.
Unladen weight	1555kg (3426lb).

Fourth Generation: The E39 Models From 1995

528i SALOON (1995 ON) AND TOURING (1996 ON)

As for 520i, except:
Engine
Cylinders	2793cc six-cylinder
Bore and stroke	84mm x 84mm with twin overhead camshafts and four valves per cylinder
Compression ratio	10.2:1
Craburretor	Bosch Motronic engine management system exhaust with catalytic converter and Lambda sensor
Max power	193bhp at 5300rpm
Max torque	206lb/ft at 3950rpm.
Tyres	225/60 R 15
Wheels	7J x 15 alloy
	16–inch and 17–inch alloy wheels optional.
Unladen weight	1440kg (3175lb).

535i SALOON (1996 ON)

As for 520i, except:
Engine
Cylinders	3498cc V8-cylinder
Bore and stroke	84mm x 78.9mm with four overhead camshafts and four valves per cylinder
Compression ratio	10.0:1
Carburettor	Bosch Motronic DME 5.2 engine management system exhaust with catalytic converter and Lambda sensor
Max power	235bhp at 5,700 rpm
Max torque	236lb/ft at 3,300 rpm.

Transmission
Gearbox	Five-speed close-ratio manual
Top	4.20:1
4th	2.49:1
3rd	1.66:1
2nd	1.24:1
1st	1.00:1
Five-speed overdrive automatic	
	3.57:1
	2.20:1
	1.51:1
	1.00:1
	0.80:1
Axle ratio	2.93:1.

Suspension and steering

Steering	Recirculating-ball type with servo assistance.
Tyres	225/55 R 16
Wheels	7J x 16 alloy wheels
	17–inch alloy wheels optional.
Unladen weight	1,615kg (3,560lb) for manual
	1,660kg (3,660lb) for automatic.

540i SALOON AND TOURING (1996 ON)

As for 520i, except:
Engine

Cylinders	4398cc V8-cylinder
Bore and stroke	92mm x 82.7mm with four overhead camshafts and four valves per cylinder
Compression ratio	10.0:1
Carburettor	Bosch Motronic DME 5.2 engine management system exhaust with catalytic converter and Lambda sensor
Max power	286bhp at 5,700 rpm
Max torque	310lb/ft at 3,900 rpm.

Transmission

Gearbox	Six-speed overdrive manual
Top	4.23:1
5th	2.52:1
4th	1.66:1
3rd	1.22:1
2nd	1.00:1
1st	0.83:1

Five-speed overdrive automatic with AGS and Steptronic control (ratios as for 535i).

Axle ratio	2.81:1.

Suspension and steering

Steering	Recirculating-ball type with servo assistance.
Tyres	225/55 R 16
Wheels	7J x 16 alloy wheels
	17–inch alloy wheels optional.
Unladen weight	1,660kg (3,660lb) for manual
	1,690 kg (3,726lb) for automatic.

Fourth Generation: The E39 Models From 1995

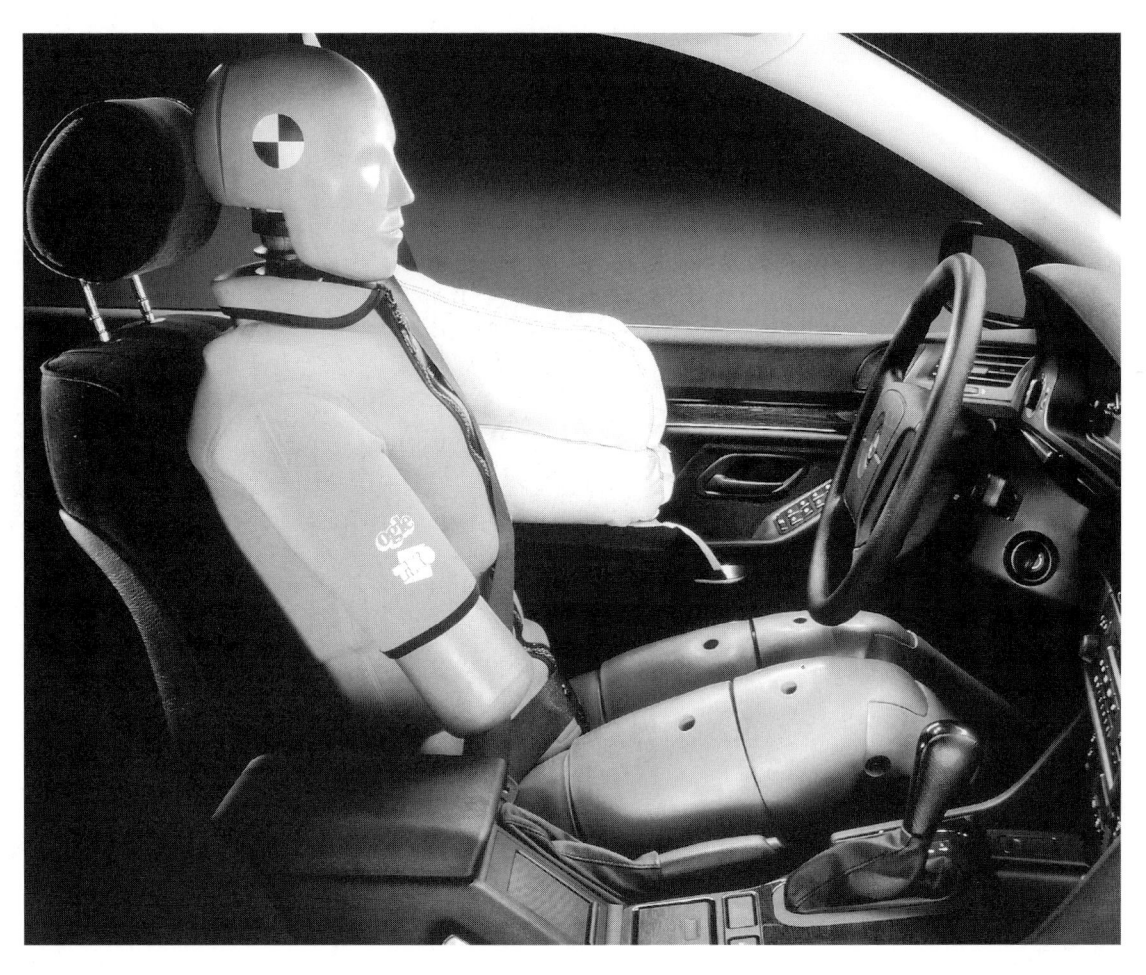

Above: Safety was an important consideration from the beginning, and these side airbags were an optional extra on the E39 5 Series cars.

the engine management system. Its wax element normally opened the circuit to the radiator only when the coolant reached 110°C, but on receipt of a signal from the engine management system, an electrically heated resistor built into the element opened the main cooling circuit below this threshold and quickly reduced coolant temperature to a safe 90°C. As a result, the engine could run hot for most of the time, but was cooled down under the sort of heavy-load conditions that would cause misfiring at those temperatures.

The engine management system was now the latest Bosch DME 5.2, seen earlier in the V12-engined 7 Series cars. It was simpler than the older type, with just one major circuit board instead of two, and it offered the one-touch starting facility previously found only on the V12 models. This meant that the ignition key did not have to be held against its spring until the engine fired; instead, a quick turn to engage the starter was interpreted by the

Fourth Generation: The E39 Models From 1995

DME system as a 'start' command, and when the key was released, the starter motor would remain engaged until the engine had fired and was running smoothly. The DME system also incorporated On-Board Diagnosis II (OBD-II), which was by this time a legal requirement in the USA and provided valuable additional diagnostic facilities for functions relevant to engine operation and emissions.

An important new transmission option was introduced at the same time as the 520i, 535i and 540i. This was known as Steptronic, and was available with the existing five-speed AGS automatic gearboxes. BMW claimed that it combined the convenience of a conventional automatic transmission with the precise and individual control of a manual unit. Steptronic provided fully automatic gearchanging in a conventional P-R-N-D gate, but alongside that (to its left in left-hand drive cars and to its right in right-hand drive models) was a second operating gate. This was marked with an M (for manual), with a plus sign above it and a minus sign below. Moving the lever forwards to the plus position against light spring pressure gave a clutchless manual upshift, and moving the lever backwards to the minus position gave a downchange. There were built-in safeguards, of course. The system ignored commands to shift down a gear if the road speed was too high, and it automatically changed up a gear just before the governed maximum engine speed if the driver failed to do so. In addition, Steptronic automatically shifted down to fourth or to third gear if the driver checked the car's

The E-39 5 Series Touring carried on from where the E34 Touring had left off, but it was a bulkier and rather less elegant design than what had gone before.

Fourth Generation: The E39 Models From 1995

speed with the brakes, so that instant acceleration was always available from a touch on the accelerator pedal.

Further new options arrived at the same time. These included leather Comfort seats with a position memory for the driver's side, sports seats, bolted cross-spoke alloy wheels and the EDC-III damper control system. Servotronic speed-sensitive steering became available right across the range, and safety could be enhanced (though, sadly, at extra cost) by side airbags in the doors, that were triggered by an impact sensor in each of the body sills.

THE TOURING MODELS, 1996

BMW's stylish Touring is perfect for the enthusiastic driver who needs more space, but will not sacrifice driving pleasure.
BMW Car, April 1997.

The major 1997 model-year news for the E39 range at the Frankfurt Show in September 1996 was that Touring derivatives were now available. The Touring had been an enormously popular derivative of the 5 Series range in its E34 incarnation, with 115,000 sales in five years, and in its best year had accounted for a quarter of all 5 Series sales. So BMW had taken the time to get the new one right – and to improve on the already impressive outgoing model. The new Touring bodyshell was made available with all the existing engine options except for that of the 535i. Right-hand drive models became available in the spring of 1997.

The basic formula was much as before in that an estate-car rear end had been grafted on to the existing saloon bodywork. Once again, the lines of the new rear section blended smoothly with those of the saloon, and the Touring looked as if it had been designed as a complete car and not as

By the time of the E39s, the US models like this one were visually indistinguishable from their European counterparts.

if it had been quickly converted from an existing saloon. As before, this had entailed redesigning the rear door window frames. However, this time the Touring models were more different from the saloons than their E34 predecessors had been.

For a start, the E39 Touring was 30mm (1.2in) longer than its saloon counterpart. The extra length had gone in behind the rear wheels in order to increase the size of the load area, and the rear suspension had also been redesigned in order to provide a flat loading floor. One of the annoying factors in so many estate cars – BMW's E34 included – was that the rear suspension towers protruded into the load area and compromised its shape and versatility. For the E39 Touring, BMW had designed a new compact rear axle with its dampers lying almost horizontally so that there was no intrusion into the load area. Pneumatic self-levelling came as standard in order to protect the handling of a heavily-laden Touring.

Under the load floor was a lockable luggage compartment of the type increasingly becoming available on estate cars, but in this case the whole of the luggage floor lifted up and could be attached to the top of the tailgate opening to provide access to this. A retractable security blind concealed the contents of the load bay, hooks and loops allowed loads to be secured without difficulty, and the rear seat was divided and could be folded forwards very easily so that it would lie flat and provide a longer load area. Options in this area of the car included a retractable luggage net to divide load area from passenger compartment. However, the most valuable option, and one that was unique to BMW and new for the E39 Touring, was an extendable loading floor. When this was fitted, the whole floor of the load area slid backwards on rails to extend beyond the tailgate by 600mm (23.6in). The advantage for loading the car was that heavy objects could be lifted onto the extended platform that could then be easily slid back into place, so there was no risk of back injury through lifting and bending into the car at the same time. In addition, the extended load floor could be used as a table for picnics or as a platform to sit on (within its weight limit of 75kg/165lb, of course), and thus added extra versatility to the car.

The E39 Touring shared the basic design of its split tailgate with the E34 Touring. So the rear window would open separately or could be treated as part of the whole tailgate. New with the E39, however, was what BMW called 'Soft Close'. This was essentially an automatic electronic closure system that pulled the tailgate shut from a near-closed position, so that it was not necessary to slam the panel shut.

The progress of the E39 range was continuous, however. After the Touring arrived, October 1996 saw the introduction of the BMW Navigation System as an option in several countries. This was based on the Philips CARiN system, and was installed in the dashboard. Essentially, it depended on a CD-ROM which carried detailed street maps, and on a satellite system which provided information about traffic conditions. By using the information displayed on the small monitor screen in the dashboard, a driver could therefore avoid traffic hold-ups ahead by choosing an alternative route. Then December 1996 brought yet another innovation in the shape of Xenon gas discharge headlamps. These gave a better controlled beam with two and a half times more light than halogen lamps, and they were also claimed to last longer – in normal circumstances, for the full life of the car. On the E39s, they were wired separately from the other electrical networks in order to ensure consistent operation.

HEAD BAGS, DSC-III, AND THE M5, 1997

For the 1998 model-year, BMW standardized yet another safety feature on the 5 Series. This was known as the Inflatable Tubular Structure (ITS), and was an airbag that spanned the window frame of each front door to protect the heads of driver and front passenger in a collision. At extra cost, buyers could order similar equipment for the rear doors. In each case, the uninflated airbag was contained within the roof lining of the car.

At the same time, the flagship 540i was fitted as standard with Dynamic Stability Control (DSC-III). This system used wheel sensors and a steering sensor to assess the safe limits of cornering. When it decided that they were about to be exceeded, it adjusted the throttle application and applied the brakes independently to each wheel in order to restore traction. In theory, therefore, it was impossible to 'lose' a 1998-model 540i through over-exuberance in a corner....

But perhaps the most exciting E39 development previewed at the 1997 Frankfurt Show was the new M5. The details of that car are explained in the following chapter.

8 The E39 Super-Saloons and Tourings

It was inevitable that the tuning specialists would get their hands on the E39 almost as soon as it reached the showrooms, and the news that BMW was not launching an M5 derivative from the start – indeed, there were stories that there might never be an E39 M5 – gave them all some valuable breathing space to develop their own performance conversions.

In many ways, however, BMW had made it hard for the tuners, and not least because the performance of the 540i was already in the super-saloon class. Perhaps

The latest generation M5 retains the looks of the standard E39 saloons, but the four tailpipes, wide tyres, and boot-lid badging make clear that this is a machine to be reckoned with.

The E39 Super-Saloons and Tourings

Beautifully presented, as ever, this is the Motorsport version of the V8 engine developed for the E39 M5. With 408bhp at 6500rpm and 369 lb/ft of torque at 3800rpm, this 4.9-litre engine would have powered the car to over 180mph if the maximum speed had not been governed to the 155mph agreed among German motor manufacturers.

the most difficult problem, however, was that the E39 was such a refined car, and that it resisted most attempts to inject it with the sound and fury that super-saloon buyers expected. Making it go faster and handle better was not a problem, but the end result often lacked that raw excitement that persuades buyers to part with large sums of money to have their cars converted in the first place.

THE HARTGE H5-2.8, 1996

The German Hartge company was the first of the tuning specialists to come up with a conversion of the E39 5 Series. The Hartge H5-2.8 was based on a 528i, and centred on a considerably uprated engine and Hartge's sports suspension. This lowered the car by 35mm and featured uprated springs and dampers, with larger-diameter

The E39 Super-Saloons and Tourings

anti-roll bars. Ultra-low-profile 245/35 ZR 19 tyres were fitted to multi-spoked alloy rims, with 8.5J wheels at the front and wider 9.5J types at the rear.

Conversion work on the 2.8-litre twin-camshaft engine started with polishing and porting the cylinder head. The valves were enlarged by 1mm and ground carefully to improve gas velocity at low lift. Their seats were enlarged, too, and reprofiled camshafts were fitted. To increase airflow, the inlet manifold was replaced by the larger-diameter item from the old 2.5-litre engine, and finally the engine management ECU was of course

Other Hartge conversions for the E39

In addition to its usual range of bodykits, alloy wheels, and suspension and brake upgrades, Hartge offered no fewer than four other engine conversions for the E39 series. The H5-2.6 was designed for the 520i, and enlarged the engine to extract 200PS. For the 528i, the H5-3.2 brought 265PS and the H5-3.5 280PS. For the 540i, the 4.7-litre engine developed for the E34 cars once again gave a Hartge H5-4.7 a power output of 340PS.

Hartge's engine conversions for the E39 ranged from a 2.6-litre 520i to a 4.7-litre 540i. Exhaust, suspension, wheels and interior could all receive the Hartge treatment, and of course there were distinctive body addenda.

reprogrammed. A sports exhaust could also be fitted as an option.

The result was a much higher-revving engine than the original, with 231bhp at 6000rpm and 221lb/ft at 4900rpm (the standard 528i had 193bhp at 5300rpm and 206lb/ft at 3950rpm). This alone was enough to transform the character of the car, and the acceleration figures told their own story. The 0-60mph time of the H5-2.8 was 6.9 seconds with manual transmission and 8.0 seconds with automatic, which compared very favourably with 7.5 seconds and 8.8 seconds for the slightly longer 0–62mph (0–100kph) sprint in the standard car.

THE ALPINA B10 V8 AND B10 V8 TOURING, 1996

... superb combination of performance, sophistication and practicality...

BMW Car, July 1997

Alpina had already developed a large-capacity V8 engine for the E34 models, and so it was a relatively straightforward operation to put that same 4.6-litre powerplant into the E39. However, in the meantime, the standard large-capacity V8 in the factory-standard 540i had grown from 4 litres to 4.4 litres in capacity, and the torque increase which came with it meant that the Alpina engine gave less of an advantage over the standard E39 540i than it had over the E34 540i. That nevertheless did not detract from the excellence of its performance, which Alpina claimed as 0–62mph (0–100kph) in 5.9 seconds with an all-out maximum of 172mph (276kph). The same figures were claimed for the slightly heavier Touring models, on which Alpina would now do their work for the first time.

The Alpina-developed V8 in 340bhp, 346lb/ft tune was only the start of the conversion, of course. The suspension of the base 540i was stiffened and lowered, and 18-inch Alpina multi-spoke alloy wheels

Hamann managed to fit the BMW V12 engine neatly into the E39, as Ian Kuah's picture shows.

The E39 Super-Saloons and Tourings

Announced in September 1997, the Hamann H5/450 V12 looked as powerful as it was.

Performance Comparison

N.B. These figures are for European-specification cars and are intended as a guide only.

Model	Power	Torque	0–62mph	Max. speed
BMW M5	408bhp	369lb/ft	5.4 sec	155mph/250kph
Alpina B10 V8	340bhp	346lb/ft	5.9 sec	172mph/277kph
Alpina B10 3.2	260bhp	243lb/ft	N/A	N/A
Hamann HM5/450 V12	450bhp	N/A	4.7 sec	193mph/311kph
Hartge H5-2.8	231bhp	221lb/ft	6.9 sec (man) 8.0 sec (auto)	N/A
Hartge 540i 4.7	340bhp	362lb/ft	5.2 sec (man) (0-60mph)	175mph/282kph
MST-Wiesmann 3.0	232bhp	237lb/ft	6.6 sec (0-60mph)	159mph/256kph
MST-Wiesmann 3.2	261bhp	249lb/ft	5.9 sec (0-60mph)	165mph/265kph
MST-Wiesmann 540i	315bhp	339.5lb/ft	N/A	N/A

were added. These were an 8J size on the front, with 235/40 ZR 18 Michelin Pilot MXX3 tyres, and a 9J size on the back, with 265/35 ZR 18 tyres of the same make. The standard braking system was also uprated, with ventilated discs and floating callipers at the rear as well as the front. As for the gearbox, the 540i's adaptive five-speed automatic was married to the steering-wheel controls of Alpina's Switchtronic to give the best of both systems.

Cosmetic addenda were limited to a straight-cut front air dam and a discreet hoop spoiler at the rear, plus badges reading 'Alpina B10 V8' on the boot lid (or tailgate). On these cars, there was no Alpina identifying badge on the grille. The dashboard of course carried the usual Alpina numbered build plate, and the upholstery was all leather, with stitching in Alpina's corporate blue and green cotton.

THE ALPINA B10 3.2, 1996

Not every customer who wanted to own an Alpina could afford the cost of a B10 V8, and to cater for those who could not, the company introduced a lower-priced model based on the mid-range 528i with five-speed manual gearbox. In this case, the engine was bored out to 86.4mm and stroked to 89.6mm, and the end result was a capacity of 3152cc. Alpina's own all-steel exhaust catalysts were fitted, together with a Boysen stainless steel exhaust featuring the visible tailpipes that Alpina customers favoured over the concealed standard type. The 3.2-litre Alpina engine gave 260bhp at 5900rpm with torque of 243lb/ft at 4300rpm, and was also used in Alpina's version of the E36 328i models, known as the B3 3.2.

The B10 3.2 had the usual Alpina suspension improvements, the same alloy

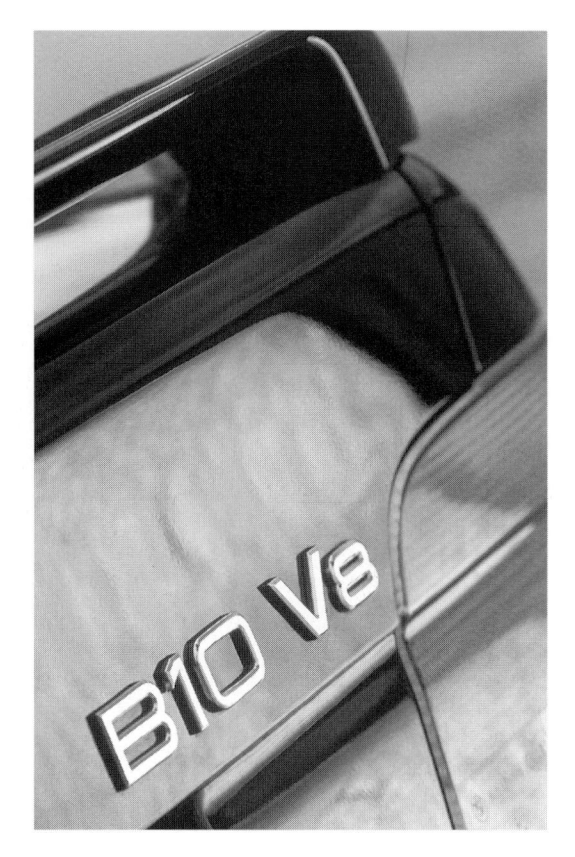

As usual, Alpina's work was accompanied by discreet badging.

wheels on low-profile tyres as the B10 V8, and the same body addenda. It was available only in the saloon models, and not in Touring form.

THE AC SCHNITZER S5S, 1996

AC Schnitzer's S5S was based on the E39 528i and offered both higher performance and eye-catching body modifications. The 2.8-litre engine was fitted with a special induction system, a modified ECU and a free-flow sports exhaust system to give 218bhp and a crisper, more sporting response.

Every Alpina has its own special manufacturer's plate. This one is on the B10 V8 pictured on p.153.

Body modifications included wider wheels and tyres, a deep front air dam, side skirts, and a choice of two different rear spoilers. Also available were a discreet roof spoiler and teardrop-shaped door mirrors. The car came with a discreet S5S badge on the rear panel.

THE MST-WIESMANN CONVERSIONS, 1996 ON

MST-Wiesmann, who had started in a small way with conversions of the E34 models, offered three engine upgrades for the E39 5 Series cars. There were 3-litre and 3.2-litre upgrades for the six-cylinder cars, and a simpler performance enhancement for the 540i.

The 3-litre engine was based on the 528i, and used a longer 89.7mm stroke to achieve a capacity of 2977cc. The same conversion could also be carried out on the 523i, although in this case its bore had to be machined out to 528i dimensions as well. The 3-litre engine had a new crankshaft and lighter pistons, and its compression ratio went up to 10.5:1. MST-Wiesmann gas-flowed the cylinder head, reshaped the valves, and fitted a high-lift camshaft with stronger valve springs. The ECU was also recalibrated and its speed limiter disabled. With 232bhp at 5300rpm and 237lb/ft at 4000rpm, the 3-litre engine gave 0–60mph in 6.6 seconds and a maximum of 159mph (256kph), plus the benefit of much sharper throttle response than either of the standard engines.

The same long-stroke crankshaft was used for the 3.2-litre conversion of the 528i. In this case, the bores were also enlarged, to 86.3mm, and a thicker head gasket was fitted. Power went up to 261bhp at 5600rpm and torque to 249lb/ft at 4000rpm, so the engine retained its low-revving character while giving 0–60mph in 5.9 seconds and a top speed of 165mph (256kph).

For the 540i, MST-Wiesmann simply reprogrammed the ECU and fitted a free-flow sports exhaust system. However, this apparently minimal work boosted the output to 315bhp at the engine's standard 5700rpm power peak and took torque up to 339.5lb/ft, again at the standard 3900rpm crankshaft speed.

To go with these performance improvements, MST-Wiesmann had of course developed suspension and braking upgrades. The suspension kit lowered the body by 40mm on the six-cylinder cars and by 30mm on the 540i. It was matched to simple but attractive 18-inch five-spoke alloy wheels of 8.5J size. On the six-cylinder cars, these wore Bridgestone Expedia S-01 tyres with the same 225/40 ZR 18 size all round, but for the 540i there

Discreet badging again – although the car itself is far from discreet! This is the Hamann H5/450 V12 pictured on page 148.

Distinctive Alpina multi-spoke alloy wheel, as used on the B10 V8.

were 235/40 ZR 18s at the front and 265/35 ZR 18s at the rear. The brakes, developed in conjunction with AP Racing, featured 360mm slotted and ventilated front discs with four-piston callipers.

The MST-Wiesmann conversions included special badging at the rear, but the company's only cosmetic offering was a deeper front air dam. Other body addenda were generally the BMW M-Technik items.

THE HARTGE 540i 4.7, 1997

Hartge took the 4.4-litre V8 of the 540i out to a larger capacity than any other tuning company in 1997 when it announced a 4,722cc version of the engine for the E39 5 Series. The engine retained its standard bore size but had a lengthened stroke, taken up to 88.8mm through the use of a new steel billet crankshaft and lighter Kolbenschmidt pistons. The compression ratio was increased to 10.5:1, and the cylinder heads were gas-flowed and fitted with valves of 1mm greater than standard diameter. The ECU was recalibrated, and a sports exhaust came as standard. The result of this work was 340bhp at the same 5900rpm as the standard engine developed its maximum 286bhp, and 362lb/ft at 3700rpm, even lower down the rev band that the standard engine's torque peak.

To make the best use of all this extra

The E39 Super-Saloons and Tourings

power and torque, the standard suspension was tightened up with Hartge's spring and damper kit which also lowered the car by 35mm. The wheels were changed for a smart spoked type of 19-inch diameter, in 8.5J size at the front and 9.5J size at the rear. To these were fitted Bridgestone S-02 tyres, 245/35 ZR 19 at the front and 265/30 ZR 19 at the rear. A 40 per cent limited-slip differential was offered as an option to prevent wheelspin when the standard ASC+T traction control was switched off.

With the standard 540i gearing and six-speed manual transmission, the Hartge 540i 4.7 could rocket to 60mph from rest in 5.2 seconds, and removal of the speed limiter from the ECU during recalibration allowed it to reach a claimed maximum of 175mph (281kph). For buyers who wanted even better acceleration, optional lower axle ratios of 3.15:1 and 3.46:1 gave even better acceleration at some cost to the car's maximum speed. The conversion was of course also availabe on a 540i fitted with the Steptronic five-speed automatic transmission.

THE HAMANN HM5/450 V12

The German tuning specialists Hamann Motorsport introduced their E39 performance conversion at the Frankfurt Show in September 1997. The car featured a long-stroke 6.1-litre derivative of the BMW V12 engine used in the 7 Series limousines, with polished and ported cylinder heads, high-lift camshafts,

The Alpina B10 V8 was similar to the later M5 in its understated visual approach.

uprated fuel injection with a reprogrammed ECU, and free-flow exhaust manifolds leading into a sports exhaust system. The result was 450bhp, a figure that gave its name to the car.

All this power was put down onto the road through an uprated clutch and six-speed manual gearbox, the latter taken from the 850CSi because the standard automatic was unable to cope with the extra torque. There was a 40 per cent limited-slip differential in the rear axle, and the chromed 18-inch wheels on lowered suspension carried ultra-low profile tyres. Hamann claimed the car's 0–62mph (0–100kph) time as an incredible 4.7 seconds, and a top speed of 193mph. There was of course a bodykit of spoilers and sills.

THE M5, 1998

The M5 derivative of the E39 saloons was a long time in coming, and in the meantime there had been a number of rumours about its specification in the motoring media. The fact that these proved surprisingly accurate strongly suggests that they had been carefully placed by BMW, to help counter the efforts of the specialist tuners while the Motorsport division completed development of the car.

The new M5 was first seen at the Frankfurt Motor Show in September 1997, but was not actually launched until the Geneva Show in the following March. As rumoured, it did become the first M5 to feature a V8 engine. This had been developed from the 4.4-litre four-valve V8, but BMW claimed that it was more than 95 per cent new. Stretched out to 4941cc and fitted with Double-VANOS (in which both inlet and exhaust camshafts offered variable timing), the new engine pumped out 408bhp at 6500rpm and 369lb/ft of torque at 3800rpm. There was a new 32-bit engine management computer, and the exhaust system had a complex manifold arrangement complemented by individual butterfly valves on each cylinder bank to smooth out the passage of the spent gases. The gearbox was an improved version of the latest six-speed manual already seen in the E36 M3, with the expected set of close ratios, and there was DSC traction control for the first time on a Motorsport-developed car. As usual, maximum speed was governed to 155mph (250kph), but BMW hinted that over 180mph (290kph) would be available if the engine were given its head.

The E39 M5 was traditionally sober in appearance. Even the engine was surmounted by a black crackle-finish intake plenum chamber, the polished aluminium example seen on the Frankfurt preview car having been abandoned for production. Nevertheless, there was no mistaking the M5 for any other E39. It sat 20mm lower on its suspension, with ultra-low profile tyres on 18-inch wheels: Dunlop 245/40ZR 18 at the front and 245/35ZR 18 at the rear. Easier giveaways were a larger front spoiler, side skirts and special mirrors, white front indicator lenses and four exhaust tailpipes in two pairs.

BMW claimed that there would be no M5 Touring this time, but it was interesting that the saloon was offered in two different states of trim – Sporting and Exclusive. The Sporting trim was self-explanatory, and was really traditional M5 fare, but the Exclusive package brought wood as well as leather, most probably to offer credible competition to Jaguar's XJR saloon. The M5's introduction to European markets was scheduled to take place over the remaining months of 1998, but no date was released for an American edition, and right-hand drive models were not expected until 1999.

9 Buying a 5 Series BMW

Despite the image associated with the BMW marque, not every one of its products is a head-turning, fire-breathing, motorway stormer. In fact, only a small proportion of the 5 Series models built in just over a quarter of a century would fit that description. The appeal of the 5 Series has always been rather different, and the factory's own standard products have always included a large element of quiet discretion in their make-up. Some of the early high-performance models undoubtedly did have some rather obvious decals and spoilers to advertise their performance potential, but BMW customers soon made clear that, on the whole, they preferred their cars without such things. So BMW left it to the tuning specialists to modify the 5 Series to suit customers who wanted to advertise their presence.

Even so, BMW's products appeal to drivers who are enthusiastic about their motoring. A small proportion of the cars probably do end up in the hands of those who see them as status symbols rather than as a beguiling blend of practical transport and tactile pleasure, but on the whole the buyer who chooses even a lowly 518i in preference to a bigger-engined but similarly-priced model from Ford, Nissan or Vauxhall is making a choice motivated by a wish to enjoy the driving pleasure which the car can give. So all of the cars covered in this book can be described as enthusiasts' cars, from the humblest to the most exotic.

Corrosion can be a problem on all BMW alloy wheels.

Alloy wheels pick up brake dust and need cleaning once in a while. These are the special TRX wheels on an E28 M535i.

Buying a 5 Series BMW

The latest 5 series engines – this is the M52 six in an E39 model – depend very heavily on electronic controls. This is likely to make DIY maintenance difficult as the cars get older.

However, the 5 Series cars have been a major success story for their manufacturer, and have sold in very large quantities worldwide. This means that, for the lesser models at least, they do not have THE rarity value of so many cars which appeal to enthusiasts. So it follows that there is no need to settle for a less than perfect example when buying second-hand: there will always be plenty more on offer, and in better condition. The exception to this rule may be the high-performance derivatives described in Chapters 4, 6 and 8 of this book. They certainly were available in far smaller numbers than the run-of-the-mill cars, and so there might be reasons for buying a well-worn example and nursing it back to health. Make no mistake, however: the job will be expensive.

So what is the appeal of these medium-sized saloons, if they are neither uncommon nor (in many cases) particularly fast? The answer is not just that they combine reliable engineering

The special metric-size tyres used with TRX wheels on the M535i of the mid-1980s are expensive to buy. Bear this in mind when buying a car which has them.

and generally excellent build quality to give long and trouble-free service: that there is a certain something about the way they drive which turns the business of getting from A to B into a pleasure. Quick throttle response, sharp steering, sporting handling (but beware of early cars in the wet!), powerful braking and an overall impression that the car is under the driver's full control are driving sensations rarely found in such generous measure in cars which are also perfectly practical everyday family transport. To drive a 5 Series BMW is to appreciate just how good an outwardly ordinary car can be. Few drivers go back to lesser makes out of choice.

Obviously, the older 5 Series cars cannot be compared with more modern cars from any manufacturer. They were excellent in their time, but car design has come a long way since the 1970s – there is no point in expecting an E12 520 to out-perform or out-handle a modern 2-litre family saloon. Even so, there is not much doubt that the BMW will still be more rewarding to drive than the more modern car. So, how should those who want to buy into the 5 Series legend go about doing so? As far as the latest E39 models are concerned, the best bet is still to buy new or to go for a dealer-warranted used car, so this chapter does not include any advice on purchasing. The earlier cars, however, will most likely be found outside dealerships and without warranties, so the advice here is designed to help intending buyers avoid the most serious pitfalls.

As a general word of advice, it is always worth buying a car fitted with a good selection of optional extras in preference to one with a more basic specification. Not only are most of these extras desirable in themselves – they include such things as sunroofs, alloy wheels and, on the later models, electronic traction aids and ABS – but they will also make the car more desirable when and if it is to be sold on. When buying, it is also worth finding out the addresses of local non-franchised BMW specialists. They will be able to assist in keeping an older car in good condition, and their prices are invariably lower than those of the franchised dealerships. Some of them may also offer cars for sale or know of customers who are thinking of selling. And, above all, they are enthusiastic about the marque. In addition, the older the car, the more worthwhile its owner will find joining one of the many enthusiasts' clubs, where like-minded owners can share problems and offer advice based on experience. Contact addresses for these clubs can be found in the classic-car magazines that are published regularly, and in specialist magazines such as Britain's *BMW Car*.

THE E12 MODELS

There were ten basic varieties of the E12 5 Series cars, of which one was available only in the USA and one only in South Africa. Three of them had four-cylinder engines, the other seven all having six-cylinder types. There were no diesels. The four-cylinder models were the 518, 520 and 520i. There was also a six-cylinder 520, and the other six-cylinder cars in most countries were the 525, 528, 528i and M535i. In the USA, only two E12 models were sold, these being the six-cylinder 530i and, later, a version of the 528i. South Africa had a six-cylinder 530 with twin-carburettor engine. A small number of cars were converted by Alpina, using either 3.3-litre six-cylinder engines or turbocharged 3-litre sixes.

BMW has justly been famous for the excellence of its six-cylinder engines, and there is no doubt that the four-cylinder types in these cars are noisier and less refined than the sixes. For high performance (0–60mph in 7.5 seconds, 136mph (219kph), the M535i is obviously the car to have, although there are not that many about. The Alpina conversions offer even better performance, although the B7 Turbo lacks the smooth power delivery so characteristic of BMWs. For a good, everyday compromise, the 528 or 528i models make an excellent choice.

All these cars offer a supple ride and, in

On high-performance cars, the brake discs take a pounding and can wear more quickly than on the average family saloon. Check condition when buying. This one is on an E28 model.

Buying a 5 Series BMW

Many enthusiastic owners would like to own one of the ultra-high-peformance models, such as an M5 or this rare Alpina B10 Bi-Turbo. The question is, can you really afford the heavy running costs, which are significantly greater than those of the standard cars?

normal conditions, excellent handling by the standards of their time. However, the trailing-arm rear suspension lets them down on wet roads, when sudden rear-end breakaway during hard cornering can be a frightening experience. Most drivers are unlikely to encounter this, but the higher the car's performance, the more likely a driver is to make use of it and to find out the hard way what happens in extreme conditions! The non-assisted steering can be rather heavy at low speeds, so it is generally preferable to look for a car with the power-assisted type.

Rust in the bodyshell was never a major problem when the cars were new, but now that some of them have been on the road for a quarter of a century or more, certain problem areas have become apparent. Rust is very commonly found on the front face of the fuel tank, which is situated just behind the right-hand rear wheel and, as a result, is constantly bombarded by stones and the like. In bad cases, the tank may leak, and rust may spread to the underbody around it. There may also be rust on the lower edges of the body just behind the front and rear wheels, and rust in the body sills and door bottoms. Corrosion can also break out around the fixing clips for the side trims.

Other problems common across the E12 range include oil leaks from the automatic transmission (usually caused by worn seals), damaged alloy wheels (that can be expensive to replace) and tyres that have worn unacceptably, either through enthusiastic driving or through neglect. Deciding which is the cause can sometimes give general pointers to other things to look for in a car!

Moving on to specific problems, the four-cylinder engines always were noisier than the sixes, but high-mileage examples develop a distinctive clatter as their camshaft wears and the timing chains begin to stretch. The only cure is replacement. As four-cylinder engines had to be worked hard to deliver the performance expected of a BMW, they are also more likely to have been 'thrashed 'than the lazier sixes. The 520i's Kugelfischer mechanical fuel injection can be a source of trouble, although the problem is usually incorrect maintenance because there is nothing fundamentally wrong with the system. Some specialists fight shy of working on the 520i simply because they do not understand its injection system.

The Bosch fuel injection, on the other hand, is very reliable if it is properly maintained. Rough running on a Bosch-injected engine is likely to be caused by nothing worse than lack of attention to filters and valves – but it does point to a low standard of maintenance in the recent past. The pre-1977 525 and 528 models can also develop carburettor troubles, as their Solex INAT types are harder to tune and lose their tune more easily than the later Solex 4A1 types. All engines seem to put wear on their water-pumps quite regularly, and the first clear indication of this is likely to be overheating. In all cases, it is also important to check that cooling systems have been filled with the correct corrosion inhibitor or anti-freeze. Alloy cylinder heads can suffer terminal damage when plain water is used as a

On earlier cars (this is an E28), look for rust where the inner and outer front wings join.

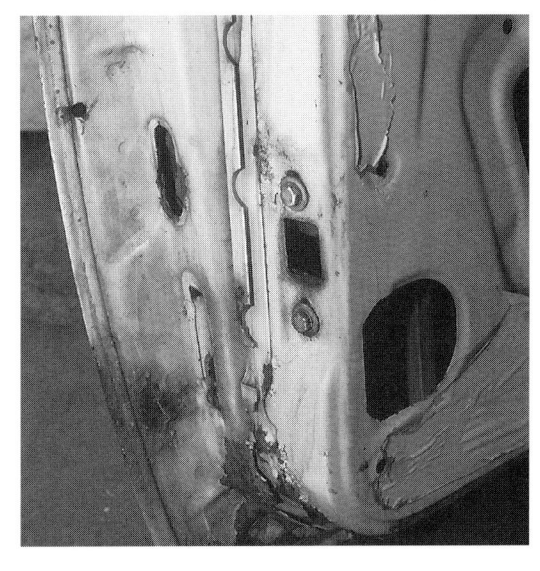

Despite generally excellent build quality, BMWs are not immune to rust, as this E28 door shows.

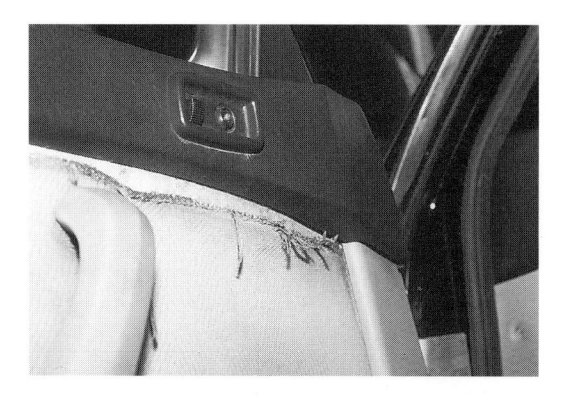

Interior problems (1): Although generally hard-wearing, door trims can show their age on older cars.

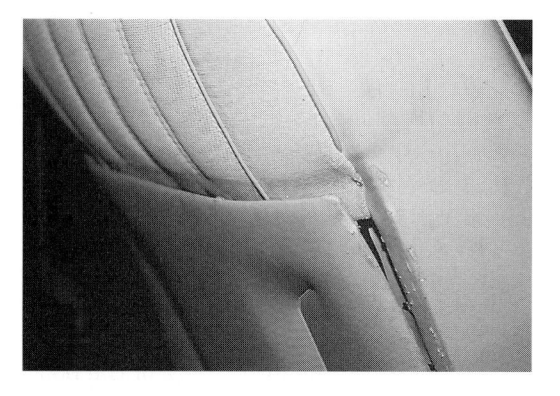

Interior problems (2): Irritating rather than catastrophic, the rear panel of this E28 seat has come adrift.

Interior problems (3): The fabric of this seat has worn badly from contact with the belt buckle. A replacement seat will probably be the only answer.

coolant; pieces of corroded alloy can also break off and block up the cooling system elsewhere, causing overheating which is almost impossible to track down.

The big six-cylinder engines (in the 525, 528, 528i, 530, 530i and M535i) suffer from the same high-mileage camshaft drive chain wear as the fours, and become similarly noisy. The problem is most obvious at idle, when there is a rattle that disappears as the revs increase. The safest course of action is always to replace the chains before they stretch enough to jump a cog on the timing wheel and cause valves to make expensive contact with pistons. Again like the fours, these engines can also suffer from worn camshaft lobes, which lead to top-end noise, poor performance and poor fuel economy.

Finally, the US-only 530i has problems all of its own. The original emissions-controlled engine lacked low-speed response and gave poor fuel economy even when new, and BMW soon modified it to quench its thirst (although the original version was still sold in California). The exhaust emissions-control system used on these engines also demanded that they run hot, and as a result BMW was faced with a plague of cracked cylinder heads when the cars were new. Many were replaced under warranty, and many more replaced under a goodwill scheme that operated for a time after the warranty had expired. However, there may still be some cars on which cylinder head cracking has yet to occur, or has not yet become obvious. It is important to know that BMW supplied a modified cylinder head after 1980, that had more metal around the water jacket and is the only one to use for replacements. The later US-model 528i, that had a different emissions control system, did not suffer from any of these troubles and is a much better buy.

THE E28 MODELS

The E28 range offers a much broader choice of models than its E12 predecessor. Outside the USA, there are two four-cylinder models (518 and 518i), two diesels (524d and 524td), and seven six-cylinder types (520i, 525e, 525i, 528i, 535i, M535i and M5). In addition, there are three US-only models (528e, 533i and 535i). On top of that, there is a greater variety of small-volume conversions from Alpina and others.

Of course everybody wants an M5, which explains why the cars are still relatively expensive. However, for everyday use, a 525i, 528i or 535i provides quite enough performance. An M535i may look faster than an ordinary 535i, but in fact it has the same performance, albeit with rather crisper handling. The 520i is a little more strained, and the engines in the four-cylinder cars have to work hard to deliver more than average performance, so reducing refinement.

The 524td and 525e deserve consideration together, because both are primarily intended to be economical cars with reasonable performance. The turbodiesel will obviously deliver better fuel mileage, while the 525e goes rather better, but is a much more relaxed cruiser than the other six-cylinder petrol cars. As for the 524d, this is a fairly rare car that was not widely available, and is probably best left doing the taxi duties for which it was designed. However, both the diesels offered astonishing levels of refinement by the standards of their day, and they should not be dismissed purely on the suspicion that they are slow and noisy.

Comfortable seats in all models and the usual supple BMW suspension give an excellent ride in all models. However, in the more sporting variants (M535i, M5), the suspension is biased towards sharper handling and the ride is therefore a little firmer. The 'sports suspension' option on other models has the same effect. The ride of the US-only 528e is also said to be marginally softer than that of other models, with consequent detrimental effects on the handling. Semi-trailing arm rear suspension can give the same tail-happy behaviour as in the older E12 models, and the modified trailing-arm arrangement in the 528i and larger-engined models simply prevents things getting out of hand too early. All the E28 models have complex electronic systems, and the higher the specification, the more complex those systems are going to be. They are not best suited to DIY repairs; serious faults are best referred to a specialist and may be costly to rectify.

Bodyshell rust is not usually serious, but the E28s do have a number of rust-prone areas. The fuel tank can suffer in exactly the same way as on the E12s, and neglected cars may have corrosion in the sills and in the double-skinned rear wheelarches, where mud can build up over a period of time. The lower edge of the boot lid may rust out, and the bottoms of the black drop-window frames can start to flake, too. A very common rust problem is in the lower A-pillar, around the hinges and check-straps for the front doors. Water collects in the door-seal retaining channels here, and will eventually set up corrosion that can cause quite alarming weakness in the pillars. Also common is rust where the inner and outer front wing panels join at the top, and more rust at the leading front corners of the outer wings. More rarely, there can be rust around the top mountings of the suspension towers in the front inner wings.

Interior trim is generally hard-wearing, but the fabric upholstery covers can wear into holes, and the only remedy for this is

Buying a 5 Series BMW

Bushes in drivetrain and suspension are critical to the refinement of all 5 Series cars.

complete replacement of the seat cover. The cloth on the door trim inserts can also shrink if it gets wet, as it will when being cleaned unless care is taken. The leather seats fitted to some models are also long-lasting, but they do need regular applications of the right sort of cleaner and of hide food to keep them supple. Cracked and dry leather can be resuscitated with care, and a few creases add a charming patina of age to these seats. However, torn leather needs to be replaced, and replacement will be costly. Problems do not generally arise with the electronically-controlled heating and ventilation system, but the Service Interval indicator can play up, most often when the batteries which operate it go flat.

The suspension and steering on these cars should have a taut feel, and any sloppiness points to wear. Hard driving accelerates wear, so worn bushes in a low-mileage car will indicate the sort of life it has led. The bushes at the ends of the front anti-roll bars wear out over time, and a rattle from the front suspension often points to this condition. At the rear, the axle beam carries large rubber-and-metal bushes, and wear here is indicated by a knocking noise when the car goes over bumps. The different rear suspension on the 528i and bigger-engined cars also suffers from wear, in this case in the Pitman joints that connect the axle ends of the trailing arms to the cross-member. As the rear of the car is subjected to the forces of acceleration and deceleration, so worn joints will move around and produce a clicking noise.

The steering and brakes are both power-assisted from the same system, so any problems with lack of assistance or poor braking may be attributable to the same fault. An engine-driven pump pressurizes hydraulic fluid, and that fluid is then stored in a hydraulic accumulator (much as on the Citroën system), and released on demand to power brakes or steering. If the accumulator or valves leak, the power assistance diminishes, so the faulty component must be replaced. On cars fitted with the all-disc braking system, there may also be trouble with the rear discs.

Where there is a choice, the five-speed overdrive manual gearboxes are better bet than the four-speed non-overdrive types. All manual gearboxes occasionally leak oil from the selector shaft seals, that are easy enough to replace. Behind the gearbox is a rubber doughnut-type universal joint, that can break up on a car which has been driven hard, and is more likely to disintegrate on one of the high-performance derivatives. Further back down the driveline, a grinding noise from under the floor when the car is being driven is usually caused by a worn propshaft centre bearing. On automatics, worn seals can also cause fluid leaks. The four-speed automatics are more responsive than the earlier three-speeds, and their overdrive top gears give better fuel consumption as well. The switchable four-speeds are probably the most desirable of all.

As for engines, the six-cylinder types are commonly good for 150,000 or 200,000 miles before they need a complete overhaul; four-cylinders are a little less tough, but also last a very long time. Anti-freeze or corrosion inhibitor is essential in the cooling system of all of them. The four-cylinder engines and the big sixes (in the 525i, 528i, M535i and M5) have chain-driven camshafts, and suffer from chain stretch and camshaft lobe wear, just like their counterparts in the E12 cars. The small-block six-cylinders (in the 520i, 525e and US-model 528e) have belt-driven camshafts, and it is important that the toothed rubber drive-belt is changed every 36,000 miles or three years, whichever is the sooner. This is an inexpensive way of avoiding the costly damage caused by a broken drive-belt. These small-block sixes can also suffer from cylinder head cracking just below the camshaft. The symptoms of this are exactly the same as for a leaking cylinder head gasket, with oil in the coolant, coolant in the oil, and excess pressure in the cooling system generally.

THE E34 MODELS

The E34 range presents an even greater variety of options than its predecessors. The saloons encompass a single four-cylinder variant (the 518i), two two-valve small-block sixes (520i and 525i), three four-valve small-block sixes (later 520i, 525i and 525iX), three turbodiesels (524td, 525td and 525tds), two two-valve big-block sixes (530i and 535i), two V8s (540i and later 530i) and two four-valve big-block sixes (3.5-litre and later 3.8-litre M5s). There are also nine Touring variants to choose from, with one four-cylinder (518i), three four-valve small sixes (520i, 525i and 525iX), two turbodiesels (525td and 525tds), two V8s (540i and later 530i) and one four-valve big six (3.8-litre M5). In addition, the tuning specialists created a whole collection of interesting high-performance models.

Sub-frame bushes must be considered as long-term consumables.

Just as with the E28s, the M5 is deservedly the most coveted of the range, and the later M5 particularly so. The two V8s are enormously sophisticated cars with a great deal of equipment as standard and big reserves of performance (although the 530i can feel a little breathless at times). The four-valve small-block sixes also offer astonishingly good performance, and are very much quicker than their two-valve predecessors, although some experts argue that they are not quite as smooth and refined. The 535i is an effortless high-speed cruiser, but the 530i brings little that the later 525i does not have. Of the diesels, the bigger-engined 525tds is the model to go for, both because of its better performance and its high equipment levels. The two-valve 520i is not as fast or as economical on petrol as most people expect, and the four-cylinder 518i is rather disappointing. Touring models of all kinds are highly prized as practical family cars, and it seems likely that their prices will remain stable long after those of the ordinary saloons have started to fall.

BMW banished rust almost entirely from the structure of the E34 range, so it is a rare car which suffers from it. Most likely, the cause will be a cheaply-executed accident repair. However, minor blemishes do show up around the nose and leading edge of the bonnet, where stone-chips that have been left untreated can eventually turn into rust. Careful owners touch such blemishes in with paint, however, so a rash of rust spots here will give a clue to the way a car has been treated.

Inevitably, the greater complexity of these cars means that there is more to go wrong on them – although in practice the additional features as compared to the earlier 5 Series ranges have a solid reliability record. One common problem is a broken or jammed sunroof operating mechanism. Another is central locking failure, particularly on post-1991 models that have a different system from that on earlier cars. The usual cause is the actuator. Service Indicator lights can give trouble, too, and just as on the E28 models the cause may be nothing more than dead batteries. Sometimes, however, the whole

Wear which can be felt in bushes may not be visible until the bushes are removed.

Service Indicator circuit board will have to be replaced, and this is a more expensive operation. Two other points are worth knowing. The first is that the later type of heated rear window (with vertical as well as horizontal bars) makes a more efficient radio aerial than the earlier type. The second is that the early ellipsoidal headlamps can be replaced by the improved super-ellipsoidal type fitted to later cars without difficulty.

Just as on the earlier 5 Series ranges, the condition of the suspension bushes is critical for the car's handling. Hard driving knocks out the front suspension bushes more quickly than average, in particular on the bottom suspension arms where these mount to the underside of the bodyshell. If the steering feels woolly or the brakes judder, these bushes should be the first thing to check. BMW did introduce longer-life bushes on later cars, and these can be fitted to earlier models. Meanwhile, at the rear, a knocking noise on acceleration or deceleration indicates worn bushes on the axle beam. In all cases, the suspension should be thoroughly checked at 100,000-mile intervals unless it has been carefully maintained in the meantime.

Looseness in the steering, which is power-assisted on all models, can be the result of worn ball-joints in the suspension, or of a worn bush on the steering idler arm. On cars fitted with the Servotronic system, some faults may also be related to the electronic control unit in the dashboard. As for brakes, the discs take a lot of punishment if these cars are used hard, and the discs themselves may need to be changed every other time the pads are renewed.

Gearboxes offer a wide variety, with five-speed overdrive, six-speed overdrive, five-speed close-ratio, four-speed overdrive automatic and five-speed overdrive automatic available at different times on different models. The taller overall gearing associated with the five-speed close-ratio manual gearboxes gives similar fuel economy to the overdrive types, so this need not affect choice; the close-ratio type does of course offer the enthusiastic driver better in-gear response if used properly. A chatter at idle from the manual gearboxes does indicate wear, but is not cause for alarm.

As for automatics, the five-speed overdrive type offers better in-gear response than the four-speed type. Fluid leaks, as on the earlier gearboxes, are mostly through deteriorated seals. Otherwise, it is worth knowing that while the four-speed automatic has a conventional dipstick for checking the fluid level, the five-speed automatic's fluid can only be checked when the car is on a garage lift. Best of the lot are undoubtedly the later automatics with AGS.

Engines are just as long-lived as on the earlier 5 Series models, as long as anti-freeze or corrosion inhibitor are used to ward off cooling system troubles. The belted camshaft drives on the two-valve small sixes also need regular maintenance attention, but on the later four-valve VANOS engines, chain drives remove the need for worry on this score. The two-valve small sixes can also suffer from cylinder head cracking if they are driven mercilessly, and their exhaust manifolds can also distort, with irritatingly noisy results. Both problems were eradicated on the four-valve versions of these engines, however. On high-mileage examples of the four-cylinder 518i engine, wear in the hydraulic tappets makes its presence felt by a rattling noise. Replacement is the remedy, but it can be expensive. The four-valve V8s are currently giving every indication of long and trouble-free service life. If they do develop problems after very high mileages, however, the sheer number of components involved is likely to make these problems expensive to put right.

EPILOGUE

BMWs in general have an undeserved reputation for being expensive to run. In 5 Series form, at least, they need not be. When things do go wrong, they can certainly be expensive to put right, but the high-quality engineering that goes into these cars means that major problems should be few and far between on cars that have been maintained to the maker's specification. That is why it is important to get some idea of how a car has been looked after before buying it. Ideally, a file full of service receipts or a BMW service book stamped in all the right places should bring

Although BMW paintwork is first-class, high-speed motoring will eventually result in stone-chips like these.

peace of mind. On earlier cars, however, that may be asking rather a lot.

One way or another, it would be foolish to buy a car that is likely to cost more than its owner can comfortably afford. The temptation to buy a high-performance model may be great, but uncomfortably high running costs and insurance may turn the whole exercise into a disaster. Equally, buying one of the smaller-engined models (particularly the 1.8-litre cars and the pre-VANOS 2-litres) may prove disappointing because the cars do not offer significantly cheaper running costs than their bigger-engined sisters, but they do offer rather less performance and refinement. The best bet, as buyers of new BMWs have known all along, is to go for one of the mid-range models. A 525i or 528i can give as much useable performance as most buyers are likely to want, and is unlikely to break the bank while doing so. In addition, a car that has been well treated by its previous owners can give many more years of practical, enjoyable and trouble-free motoring, no matter how old it may be.

Index

AC Schnitzer and E34 models 133
    E39 S5S 171
Adaptive M-Technic suspension 137
AIC rain-sensor wipers 148
Airbags, Euro type 126
Alpina:
    E12 B7 Turbo 32
    E28 B7S 84
    E28 B9 3.5 87
    E28 B2.8 88
    E28 B10 95
    E34 B10 117
    E34 B10 Bi-Turbo 134
    E34 B10 4.6 140
    E39 B10 V8 168
    E39 B10 3.2 170
ASC wheelspin control 103
ASC + T traction control 147

Bangle, Chris 143
Baroque Angels 9, 13
Bertone 23, 25
BMW, crisis and rebirth in the 1950s 16
BMW in the post-war period 12
BMW in the USA 30
BMW models (other than 5-series):
    02 range 19
    503 16
    507 19
    600 15
    700 15
    Baroque Angels 9, 13
    Neue Klasse 18
Bracq, Paul 23
Buying a used E12 178
    E28 183
    E34 185

Cairo assembly plant 147

Catalytic converters and Germany 78

DDE (Digital Diesel Electronics) 102
DDE II 123
DME (Digital Motor Electronics) 160
Dechroming option 126
Diesel engine history 54
Dynamic Stability Control (DSC-III) 164

E-series codes 8
EDC-III damper control 147
Electronic Check Control, in E34 99
Engines, in E28 models 54
    in E34 models 101
    in E39 models 149
EWS immobiliser 109
EWS-II system 148

Family Touring prototype (E34) 109
Fiedler, Fritz 18
5-series models:
    518 (E12) 33
    518 (E28) 53
    518g Touring (E34) 107
    518i (E28) 76
    518i (E34) 102
    518i, revised (E34) 126
    520 four-cyl (E12) 29
    520 six-cyl (E12) 31
    520i (E12) 29
    520i (E28) 56
    520i (E34) 101
    520i 4-valve (E34) 105
    520i (E39) 149
    523i (E39) 149
    524d (E28) 83
    524td (E34) 101
    525 (E12) 31
    525e (E28) 62

Index

525i (E28) 76
525i (E34) 99
525i 4-valve (E34) 105
525iX (E34) 110
525td (E34) 126
525tds (E34) 113
525tds (E39) 149
528 (E12) 36
528i (E12) 47
528i (E28) 56
528i (E39) 149
530i (E34) 101
535i (E28) 78
535i (E34) 101
535i (E39) 152
M535i (E12) 49
M535i (E28) 78
535i (E34) 101
540i (E34) 123
540i (E39) 152
M5 (E28) 53, 91
M5 3.5-litre (E34) 128
M5 3.8-litre (E34) 136
M5 (E39) 165
Touring (E34) 107
Touring (E39) 162

Garmisch prototype 25
Goertz, Albrecht 19

Hahnemann, Paul 28
Hamann HM5/450 V12 174
Hartge:
    E28 H5S 90
    E34 conversions 139
    E39 H5-2.8 166
    Other E39 models 167
    E39 540i 4.7 173
Heidegger (Liechtenstein BMW agent) 32
Hoffmann, Max 19
Hofmeister, Wilhelm 19

Inflatable Tubular Structure (ITS) 164
Isetta 11

Janspeed twin turbo E34 535i 136

Lancia, failed purchase 22
Lange, Karlheinz 60
Lincoln, BMW diesel engine in 61

M5 convertible prototype 128
Motronic engine management 77
MST-Wiesmann E34 conversions 140
    E39 conversions 166

Name, choice of, for 5-series 28
Natural gas E34 102
Navigation system 163
Neue Klasse saloons 18

On-Board Diagnostics (OBD-II) system 161

Performance comparisons:
    E28 super-saloons 92
    E34 super-saloons 141
    E39 models 148
    E39 super-saloons 169
Production of E34 models 105
Production of E39 models 143

Quandt, Harald and Herbert 18

Radermacher, Karlheinz 60
Recession, effects of 108
Recycling and BMW 111
Rosslyn factory 34, 79

SE models, in UK (E34) 110
Security, in E34 range 109
Servotronic steering 102
South African variants:
    E12 518i 52
    E12 530 34
    E28 M535i 78
    E28 M5 94
    E34 range 105
Sport models, in UK (E34) 110
Steptronic transmission control 156

Index

Sytner, UK Alpina agents 88

Technical specifications tables:
    E12 38
    E28 64
    E28 M5 88
    E34 114
    E34 M5 3.5-litre 131
    E34 M5 3.8-litre 136
    E39 156
TWR-tuned E28 535i 59

US variants:
    E12 528i 48
    E12 530i 36
    E28 528e 70
    E28 533i 71
    E28 535i 81
    E28 M5 94
    E34 540i Sport 103

V8 engines 123
VANOS engines 124
Von Falkenhausen, Alex 19

Wolff, Eberhard 18

Xenon gas discharge headlamps 163